MULTI-STANDARD CMOS WIRELESS RECEIVERS: Analysis and Design

THE KLUWER INTERNATIONAL SERIES IN ENGINEERING AND COMPUTER SCIENCE

ANALOG CIRCUITS AND SIGNAL PROCESSING

Consulting Editor*: Mohammed Ismail. *Ohio State University

Related Titles:

DATA CONVERTERS FOR WIRELESS STANDARDS
C. Shi and M. Ismail
ISBN: 0-7923-7623-4
AUTOMATIC CALIBRATION OF MODULATED FREQUENCY SYNTHESIZERS
D. McMahill
ISBN: 0-7923-7589-0
MODEL ENGINEERING IN MIXED-SIGNAL CIRCUIT DESIGN
S. Huss
ISBN: 0-7923-7598-X
CONTINUOUS-TIME SIGMA-DELTA MODULATION FOR A/D CONVERSION IN RADIO RECEIVERS
L. Breems, J.H. Huijsing
ISBN: 0-7923-7492-4
DIRECT DIGITAL SYNTHESIZERS: THEORY, DESIGN AND APPLICATIONS
J. Vankka, K. Halonen
ISBN: 0-7923 7366-9
SYSTEMATIC DESIGN FOR OPTIMISATION OF PIPELINED ADCs
J. Goes, J.C. Vital, J. Franca
ISBN: 0-7923-7291-3
OPERATIONAL AMPLIFIERS: Theory and Design
J. Huijsing
ISBN: 0-7923-7284-0
HIGH-PERFORMANCE HARMONIC OSCILLATORS AND BANDGAP REFERENCES
A. van Staveren, C.J.M. Verhoeven, A.H.M. van Roermund
ISBN: 0-7923-7283-2
HIGH SPEED A/D CONVERTERS: Understanding Data Converters Through SPICE
A. Moscovici
ISBN: 0-7923-7276-X
ANALOG TEST SIGNAL GENERATION USING PERIODIC ΣΔ-ENCODED DATA STREAMS
B. Dufort, G.W. Roberts
ISBN: 0-7923-7211-5
HIGH-ACCURACY CMOS SMART TEMPERATURE SENSORS
A. Bakker, J. Huijsing
ISBN: 0-7923-7217-4
DESIGN, SIMULATION AND APPLICATIONS OF INDUCTORS AND TRANSFORMERS FOR Si RF ICs
A.M. Niknejad, R.G. Meyer
ISBN: 0-7923-7986-1
SWITCHED-CURRENT SIGNAL PROCESSING AND A/D CONVERSION CIRCUITS: DESIGN AND IMPLEMENTATION
B.E. Jonsson
ISBN: 0-7923-7871-7
RESEARCH PERSPECTIVES ON DYNAMIC TRANSLINEAR AND LOG-DOMAIN CIRCUITS
W.A. Serdijn, J. Mulder
ISBN: 0-7923-7811-3
CMOS DATA CONVERTERS FOR COMMUNICATIONS
M. Gustavsson, J. Wikner, N. Tan
ISBN: 0-7923-7780-X
DESIGN AND ANALYSIS OF INTEGRATOR-BASED LOG -DOMAIN FILTER CIRCUITS
G.W. Roberts, V. W. Leung
ISBN: 0-7923-8699-X
VISION CHIPS
A. Moini
ISBN: 0-7923-8664-7

MULTI-STANDARD CMOS WIRELESS RECEIVERS: Analysis and Design

XIAOPENG LI
Data Converter Department
Texas Instruments Incorporated
Dallas, TX 75243
USA

MOHAMMED ISMAIL
Analog VLSI Laboratory
The Ohio State University
Columbus, OH 43210
USA

Springer Science+Business Media, LLC

ISBN 978-1-4757-7602-7 ISBN 978-0-306-47309-8 (eBook)
DOI 10.1007/978-0-306-47309-8

Library of Congress Cataloging-in-Publication Data

Originally published by Kluwer Academic Publishers in 2002
Softcover reprint of the hardcover 1st edition 2002

Printed on acid-free paper.

Contents

List of Figures

List of Tables

Preface

Wireless communication is witnessing tremendous growth with proliferation of different standards covering wide, local and personal area networks (WAN, LAN and PAN). The current trends call for designs that allow:

- Smooth migration to future generations of wireless standards with higher data rates for multimedia applications;
- Convergence of wireless services allowing access to different standards from the same wireless device;
- Cost effective design solutions for intercontinental roaming.

This requires designs that work across multiple wireless standards, can easily be reused, achieve maximum hardware share at minimum power consumption levels particularly for mobile battery-operated devices.

While serious efforts are currently underway to develop highly integrated solutions for digital basebands covering multiple standards, today's emerging multi-standard, multi-band wireless devices use "stacked" transceivers, i.e. separate transceivers for different standards. This represents a major bottleneck in attempting to achieve higher levels of integration and reduce the bill of material for a multi-standard wireless device. Development of radio architectures and mixed-signal design solutions that support multiple standards is therefore needed. This is one of the main objectives of this book.

This is the first book on the subject of multi-standard wireless receivers. It covers both the analysis and design aspects of CMOS radio receivers with primary focus on receivers for mobile terminals. The subject of multi-standard data converter design for base stations is also covered. The book is the first to detail a complete analysis of a multi-standard receiver chain from the antenna to the data converter establishing the design specifications of all blocks in both the radio and mixed-signal parts. The specifications are conveniently tabulated throughout the book and span a level of details from the noise figure and linearity requirements of RF blocks down to such circuit details as, e.g., DC gain, settling behavior, noise and matching requirements of amplifiers (OTAs) in the data converter. The focus is on the third- generation backward-compatible UMTS (WCDMA and GSM standards) as well as the short distance cordless DECT standard. However, the techniques and design methods presented in the book can easily be adapted to other multi-standard application scenarios encompassing, for example, GPS, WLAN and Bluetooth.

As such the book will serve as a reference for RF IC and mixed signal designers, wireless receiver system designers, design managers and project leaders in industry, particularly those in the wireless semiconductor industry. The book could also serve as a reference or a text for a first year graduate course on the subject for electrical and /or computer engineering majors.

One of the most critical challenges to overcome in the design of multi-standard terminals for mobile and personal communication systems is to find the maximum number of commonalties amongst the different wireless standards of interest to allow the highest possible degree of integration, under the constraints of low voltage and low power consumption. The starting point in the design process of multi-standard terminals is to set up a common system architecture capable of implementing various standards and supporting higher levels of system integration of both the radio and the digital baseband parts. This involves a system partitioning strategy leading to a chip set, a system in package or ultimately a single system-on-chip solution. This also calls for radio design solutions with mixed signal strategies that take full advantage of CMOS technology scale-down by moving functions, such as channel select filtering, offset compensation, modulation and demodulation, to the digital domain.

After a brief introduction of multi-standard wireless applications in Chapter 1, the book presents the basic receiver architectures and system level design principles in Chapter 2. In Chapter 3, a multi-standard Zero-IF receiver for WCDMA/DCS1800/DECT is discussed at the system level. A top-down CMOS-specific mapping strategy of system-to-block level specifications is adopted. This is based on knowledge of what performance levels can realistically be achieved when performing design at the block level in present-day deep sub-micron CMOS technologies. Moreover, a strategy for the allocation of filtering specifications between the baseband analog filter and analog-to-digital converter (A/D) is developed to maximize hardware sharing and minimize power consumption. The detailed receiver chain from the antenna to the A/D converter is analyzed. Multi-standard RF front-end and synthesizer subsystems are then discussed in Chapter 4. Novel RF CMOS design techniques for LNAs and prescalers are presented.

A programmable $\Delta\Sigma$ modulator A/D converter for multi-standard receivers is introduced in Chapter 5, together with discussions of architecture optimization, circuit nonideality analysis and simulation. Chapter 6 presents the circuit design of the various building blocks of the A/D converter.

In contrast with the wireless handset, the wireless base station receiver presents a different set of challenges from the design point of view. Here performance is more important than power consumption. High-speed high-resolution A/D converter is an obvious bottleneck in a multi-standard, multi-carrier base station application. In Chapter 7, a novel technique for high-speed high-resolution A/D converter is developed, which provides high-speed calibration of all nonidealities in pipeline architectures.

This book has its roots in the doctoral dissertation work of the first author at the Analog VLSI Lab, The Ohio State University. We would like to thank all those who supported us at the Analog VLSI Lab and at other locations including the Lab of Electronics and Communication Systems (LECS) at the Swedish Royal Institute of Technology, The Data Converter Group at Texas Instruments Corporation, Dallas and Spirea AB, Stockholm.

XIAOPENG LI AND MOHAMMED ISMAIL

COLUMBUS, OHIO

2002

This book is dedicated to Li Guozhang, Zhang Cuimei, and Sameha, Ismail Sr., Tuula, Ismail Jr. and Omar

Chapter 1

WHY MULTI-STANDARD RECEIVERS

1. Introduction

Since the initial commercial introduction of the advanced mobile phone system (AMPS) service in 1983, mobile communication has seen an explosive growth worldwide [56]. Besides the frequency reuse capabilities provided by the cellular operation, advances in technologies for wireless access, digital signal processing, integrated circuits, and increased battery life have continued the exponential growth in mobile and personal communication services. Systems are cellular or wide area networks (WAN), cordless telephony, satellite mobile, paging and other specialized mobile radio systems including wireless system for local, personal and metropolitan area networks (LAN, PAN and MAN). Whereas the analog cellular mobile systems fall in the category of first-generation mobile systems, the digital cellular, low power wireless, and personal communication systems are perceived as second-generation mobile systems. The emerging Internet environment urgently requires support for asymmetric, interactive, multimedia traffic based on high-speed packet data transport. Such rapidly growing service requirements, driven by the global users of telecommunications, will dramatically change the nature of telecommunication services in the twenty-first century. The underlying vision for the emerging mobile and personal communication services for the new century is to enable communication with a person, at any time, at any place, and in any form, with a paradigm shift from the current focus of voice and low speed data services to high speed data and multimedia services.

The current second-generation digital mobile and personal communication systems are based on national or regional standards that are optimized for region- or country-specific regulatory and operating environments. They are therefore unable to interoperate with each other and can provide mobility only with their radio environments as well as within geographic regions in which a specific standard is operational. Efforts are therefore under way at the international as well as the regional/national levels to develop the so-called third-generation mobile telecommunication system that will meet the coming needs of telecommunications subscribers.

2. Digital Cellular Mobile Systems

Currently there are four major digital cellular standards, GSM, IS-136 (or D-AMPS), PDC and IS-95. Except for IS-95, which is a CDMA technology, all the other three are based on TDMA technology. The air interfaces of these standards are included in Table 1.1. GSM is a standard that was developed by the European standards committee. The original version of GSM was used in the 900MHz band throughout Europe. Then an upbanded version of GSM was added in the 1800MHz band, this was originally called PCN for personal communications network. PCN was later renamed DCS1800. An extra 10MHz was added to the lower end of both receive and transmit bands to create Enhanced-GSM or E-GSM. The European GSM is currently being deployed in the United States in the 1900MHz band and is named PCS1900. GSM and DCS1800 have been adopted by a large number of operators worldwide and have captured the largest global subscriber base among current digital cellular mobile systems.

Table 1.1. Summary of Cellular Phone, Cordless Phone, LAN and PCS Standards

Wireless Standard	Access Scheme	Frequency Specturm (MHz)	Channel Spacing	Frequency Accuracy	Modulation Technique	Data Rate	Peak Power
AMPS	FDD	824 849(Tx) 869 894(Rx)	30kHz	2.5ppm	FM	N/A	3W
DCS-1800	TDMA	1710 1785(Tx) 1805 1850(Rx)	200kHz	90Hz	GMSK	270.8kbs	0.8,2, 5,8W
GSM	TDMA/ FDMA/ FDD	890 915(Tx) 935 960(Rx)	200kHz	90Hz	GMSK	270.8kbs	0.8,2, 5,8W
EGSM	TDMA	880 915(Tx) 925 960(Rx)	200kHz	90Hz	GMSK	270.8kbs	0.8,2, 5,8W
PCS-1900	TDMA	1880 1910(Tx) 1930 1930(Rx)	200kHz	90Hz	GMSK	270.8kbs	0.8,2, 5,8W
IS-54 (IS-136) (D-AMPS)	TDMA/ FDD	824 849(Tx) 869 894(Rx)	30kHz	200Hz	/4 QPSK	48kbs	0.8,1, 2,3 W
DECT	TDMA TDD	1881 1897	1.728MHz	50kHz	GFSK	1.152Mbs	250mW
802.11 (DSSS)	CDMA	2400 2483	N/A	25ppm	QPSK	1,2,11Mbs	1W
WCDMA (UMTS)	CDMA	1920~1980(Tx) 2110~2170(Rx)	5MHz	+/- 0.1ppm	QPSK	3.84Mbs	0.125, 0.25,0.5 2W
IS-95	CDMA	824-849(Tx) 869-894(Rx)	1.25MHz	N/A	OQPSK	1.2288Mbs	
Bluetooth (802.11FH)	CDMA/ FH	2400-2483	1MHz	20ppm	GFSK	1Mbs	0-20dBm

In GSM the uplink (mobile-to-base) frequency band is 890-915MHz and the corresponding downlink (base-to-mobile) band is 935-960MHz, resulting in a 45 MHz spacing for duplex operation. The GSM uses time division multiple access (TDMA) and frequency division multiple access (FDMA), whereby the available 25MHz spectrum is partitioned into 124 carriers, with the carrier spacing of 200 KHz, and each carrier in turn is divided into 8 time slots. Each slot is one radio channel. Each user transmits periodically in every eighth time slot in an uplink radio carrier and receives in a corresponding time slot on the downlink carrier. Thus several conversations can take place simultaneously at the same pair of transmit/receive radio frequencies.

3. Low Power Wireless Communication Systems

Digital cordless telecommunication systems are intended to provide terminal mobility in residential, business, and pubic access applications where users can originate and receive calls on their portable terminals as they change locations and move about the coverage area at pedestrian speeds. It is also anticipated that the same terminal can be used in the three application environments: at the residence, at the workplace, and at such public locations as airports, train and bus stations, and shopping centers.

In contrast to cellular radio, cordless telecommunications standards primarily offer an access technology rather than fully specified radio access and network standards. Cordless terminals generally transmit at lower power than cellular, resulting in the use of microcells. In high density (in-building) applications much smaller cells (picocells) may be used so that significantly higher traffic densities can be supported. Furthermore, cellular networks generally operate in an environment characterized by regulation (which may vary from country to country), a limited number of operators, centrally managed frequency resources, and relatively high infrastructure costs. Cordless telecommunication systems, on the other hand, are expected to operate in an unregulated, open market environment, where system installation and frequency planning can not be coordinated or planned, and cost and performance as perceived by the end user are key factors in market acceptance.

The cordless telephony standards adopted by European standards are CT2 (Cordless Telephony 2) and DECT (Digital Enhanced Cordless Telecommunications). PHS (Personal Handy-phone System) was standardized in Japan, and PACS (Personal Access Communication System) is a North American low power PCS standard. In Europe the 1880-1900MHz band has been set aside for DECT. To utilize the available 20MHz band in an efficient and flexible manner for supporting voice and data applications, the DECT standard provides for space, frequency, and time distribution. Space dispersion in DECT is supported through a frequency reuse feature based on the cellular concept. To provide frequency distribution, the available spectrum is segmented into

10 carrier frequencies (frequency channels) from 1881.792 to 1897.344 MHz with a separation of 1.728MHz. Time distribution is achieved by using time division multiple access (TDMA) and time division duplex (TDD) methods, whereby each frequency channel supports 12 duplex time slots or 32kb/s channels. Thus a total of 120 channels, each with a 32 kb/s data rate, are available in the FDMA/TDMA/TDD structure to carry voice and data traffic.

Besides the application of DECT in the residential, small and large business, and public access (telepoint) environments, DECT can provide additional capacity and coverage to GSM service providers. The dual-mode terminals deployed in the interworking architecture allow the users to switch to the DECT system when they are in the DECT coverage area. There has been considerable interest in the concept of dual mode GSM/DECT phones. Such phones allow the user to use the wide-area GSM cellular network while out of the office or away from home, also using the same handset at home or via the office PABX when in range of a DECT base station - "one phone anywhere". At present there is little available in the way of commercial dual mode telephone products. Some believe that DECT/GSM dual mode phones will be an important niche market in the coming years, but differ as to when the market will take off.

4. Third-Generation Mobile Communication Systems

The international Telecommunication Union (ITU), the United Nations organization responsible for global telecommunications standards, has been working since 1986 toward developing an international standard for wireless access to worldwide telecommunication infrastructure. This standard is known as IMT-2000, for International Mobile Telecommunications 2000, where 2000 indicates the target availability date (year 2000) as well as the operational radio frequency band (2000 MHz range). IMT-2000 is intended to form the basis for third-generation (3G) wireless systems, which will consolidate today's diverse and incompatible mobile environments into a seamless radio and network infrastructure capable of offering a wide range of telecommunication services on a global scale, shown in Figure 1.1. IMT-2000 will provide capabilities constituting significant improvements over the current mobile systems, especially in terms of global mobility for the users and support of services like high-speed data, multimedia, and Internet. Since, however, it is generally accepted that these IMT-2000 capabilities will to a large extent be achieved by evolving existing wireless and wireless networks, IMT-2000 will be a family of systems rather than a single, monolithic network.

WCDMA (Wideband Code-Division Multiple-Access) has been chosen as the basic radio access technology in Europe and Japan for wideband radio access to support third-generation multimedia services [23] [36]. Optimized to allow very high-speed multimedia services such as voice, Internet access and videoconferencing, the technology will provide access speeds at up to 2Mbit/s

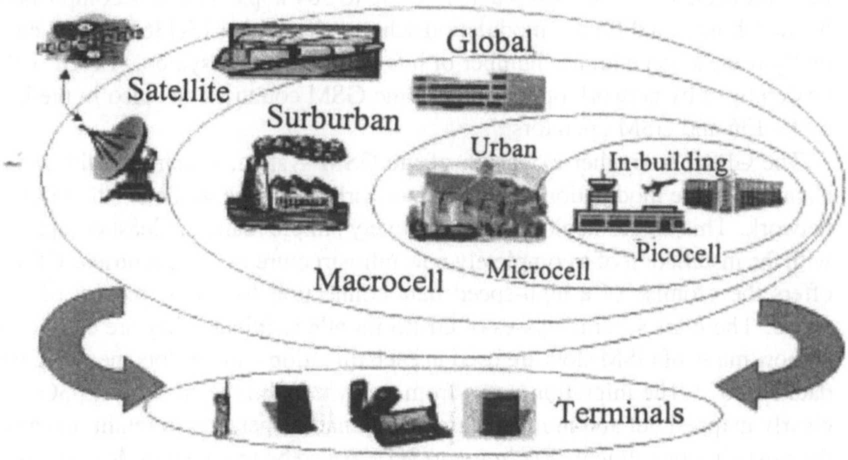

Figure 1.1. World All-in-one Phone Scheme

in the local area and 384kbit/s wide area access with full mobility. These higher data rates require a wide radio frequency band, which is why WCDMA with 5MHz has been selected; compared with 200kHz for narrowband GSM. Compared to second-generation narrow-band CDMA, the WCDMA radio interface offers significant improvements, in addition to the support of higher date rate services. These include: improved coverage and capacity due to a higher bandwidth and coherent uplink detection; support of inter-frequency handover; support for capacity-improving technologies such as adaptive antennas, multi-user detection; and a fast and efficient packet-access protocol [13].

WCDMA radio interface also includes the circuit-switched mobile services switching center of GSM and the packet-switched GPRS (General Packet Radio Service) support node. Thus WCDMA can be added to existing GSM core networks. This will be particularly beneficial when large portions of new spectrum are made available, for example in the new 2GHz bands in Europe and Asia. It will also minimize the investment required for WCDMA infrastructure as it will be possible for existing GSM sites and equipment to be reused to a large extent.

5. Migration from Second to Third Generation

The so-called "2.5G" systems represent intermediate upgrade of the 2G system in data rates available to mobile users. While there is no clear consensus of what constitutes 2.5G, a few systems certainly qualify. Enhanced Data GSM Evolution (EDGE) is a representative 2.5G system. It uses the GSM or IS-136 network backbone, and allows data rates up to 384 kbps. This is accomplished by switching to a different modulation scheme than either IS-136 or GSM currently uses, to get a higher number of bits per transmitted symbol. EDGE will be deployed by network operators in some GSM countries and also in the US by IS-136 and GSM operators.

The GPRS is another extension of the GSM system, which uses the same channels, same modulation, and same network backbone as the existing GSM network. This makes deployment a relatively simple matter, at least compared with the installation of a completely new infrastructure in 140 countries. GPRS offers the promise of a high-speed data connection for users in most of the world. There are several classes of GPRS mobile terminal. They are classified by how many of GSM slots are used in each direction – more slots mean higher data speed. The migration paths from today's mobile technology to 3G are clearly mapped out and in many cases, will enable operators to retain much of their existing investment. Evolution to 3G can also be progressive, so operators can control just how fast - or slow - migration of subscribers takes place.

Together, GPRS and EDGE mean that operators with 2G networks today can evolve to 3G services on existing network frequencies, using current network infrastructure. This means that operators do not need additional higher frequency bands and can simply build on their existing investments, through some relatively straightforward hardware and software investments. New handsets will be needed to support both GPRS and EDGE. In fact, GPRS handsets have already started to appear in the market (e.g. the Nokia 6310 and the Ericsson T39 with triple-band, GPRS and Bluetooth).

It is also important to note that increased licensing costs for 3G WAN has motivated increased development of short distance standards like LAN and PAN using the unlicensed ISM bands of 2.4 and 5GHz, where access to the internet for voice over IP and other mobile internet applications is possible. Also, in hot spots, e.g. shopping malls, airport lounges, hospitals, etc, where capacity is limited, access to WAN can be achieved through LAN access points. Such combination is referred to as metropolitan area network (MAN).

6. Software Radio and Multi-Standard Receivers

The Software Radio (SWR) is such a radio with functions that may be redefined in software (e.g., by downloads or subscriber identity cards). This brings high potential for service differentiation. Early SWR research focused on the military's goal of total flexibility across dozens of radio frequency (RF) Bands. Current research focuses more on economic viability in introducing third-generation (3G) handsets and infrastructure. An ideal SWR is a multi-band multi-mode radio with dynamic capability defined through software in all layers of the protocol stack, including the physical layer, which is the subject of this book. In base stations, a wide-band analog-to-digital converter (ADC) may convert an entire service band at IF, with programmable digital filters for channel isolation and DSP's for demodulation. Coupled with a corresponding wideband digital-to-analog converter (DAC) and software-reprogrammable functions from baseband to IF, such a radio supports arbitrary air interfaces within engineering limits. These limits include the bandwidth and dynamic range of the ADC and DAC, timing accuracy, and the processing capacity of the digital processing hardware (including reprogrammable ASIC, FPGAs, DSP chips, and general-purpose processors). Obviously, the approaches to software radio handsets and software radio base-station can be quite different. The base-station needs to process all the signals in the service band while power consumption is not a major issue. On the contrary, battery-operated mobile handsets are only required to select and process one channel while the power consumption must be kept low. Thus the lack of ADC performance, high-speed and high-resolution, is the economic showstopper for base station SWRs. The state-of-the-art 14bit 70MHz pipelined ADC is even not enough for digital-IF GSM basestation receivers. There are more challenges on the handsets side. The challenges for software radio handsets are shown in Table 1.2 based on investigation of present and future mobile communication systems.

Table 1.2. SWR and Multi-Standard implementation challenges

Challenge area	Challenges	Approaches
Power management	Sleep vs. paging delay	Power-managed DSP devices
Clocking	Multiple standards	Normalization, low power
Receiver architecture	Efficiency vs. Multiple standards	Direct conversion or wideband SWR
Parts count	Off-chip passive devices	MEMS
Software	Computational efficiency	FPGAs, Java engines

7. Summary

The software radio receiver or multi-standard receiver is essential for future all-in-one cellular phones. What standards should be included in the receiver differs for different areas. For example, in North America, it is more reasonable to develop a multi-standard receiver covering AMPS/IS-95/cdma-2000. In this book, we focus on the European standards. Moreover, as mentioned earlier, it is necessary and beneficial to integrate cordless phones, like DECT, together with cellular standard, which leads to a multi-standard receiver covering WCDMA/GSM/DECT. As it is shown, there are many challenges for software radio receivers at each level or layer. From the viewpoint of a RF front-end receiver designer, whose job is to complete the physical layer or air interface protocol, the challenge is to design a receiver that supports all future standards with backward compatibility, from 2G, 2.5G to 3G system, from cellular, cordless to satellite mobile. Support to short distance standards, i.e. PAN and LAN, will also be required. This book demonstrates the feasibility of such a mobile terminal with low power, highly efficient fully integrated single-chip realization. In this work, we chose the most popular standards, GSM (DCS1800), DECT and WCDMA to be integrated on the same receiver chip. The methods, architecture and techniques can be easily extracted to implement other types of multi-standard integrated receivers, e.g. a design combining Bluetooth with WLAN 802.11b and/or HomeRF in the 2.4GHz ISM band. For different specifications, the actual realization may be different. However, note for example, that the 2.5G standards, such as EDGE and GPRS, have similar air interface as GSM. Thus the GSM RF front-end receiver design inherently supports EDGE and GPRS standards.

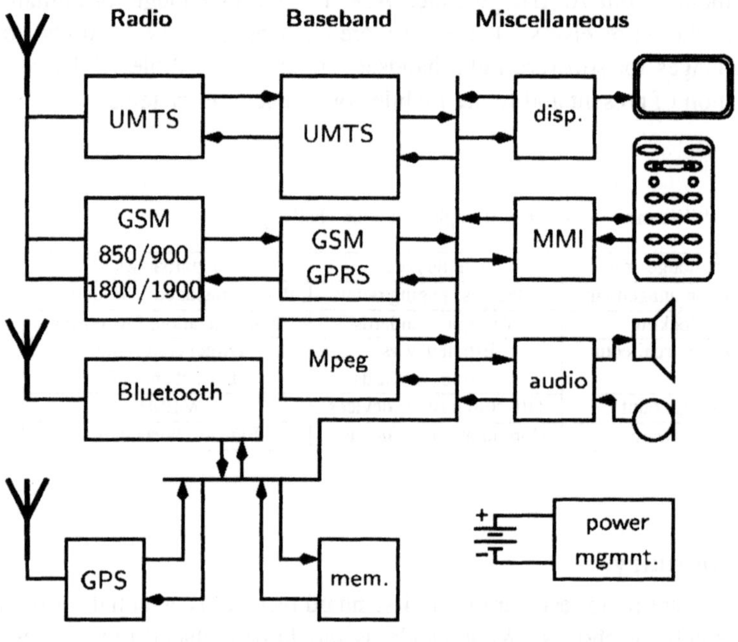

Figure 1.2. Bluetooth-enabled Multi-Standard Handset [50]

An example of a wireless system of a Bluetooth- and GPS-enabled 3G handset is shown in Figure 1.2 [50]. A serious effort is currently under way by companies in the wireless semiconductor business to integrate baseband parts into a single digital system-on-chip (SOC). This however is beyond the scope of this book. In Figure 1.2, "stacked" separate radio transceivers are used for different standards. In contrast, our objective is to show that a fully integrated radio architecture can support multiple standards.

The latter part of the book focuses on high-speed high-resolution A/D converter techniques, which is the bottleneck for base-station software radio receivers. Though in this part, the application is not specifically defined, the fast calibration algorithm discussion for pipelined A/D converters is essential for achieving high-performance software radio receivers.

Chapter 2

RECEIVER ARCHITECTURES AND DESIGN BASICS

The specifications of a radio receiver are determined by the system parameters. The system parameters include operating frequency band, blocking characteristics, intermodulation characteristics, sensitivity level, adjacent channel rejection and etc. This chapter reviews the fundamental challenges and trade-offs in radio receiver design. The basic concepts and equations discussed in this chapter are required in Chapter 3, where the proposed multi-standard receiver specifications are analyzed and distributed to building blocks. Different receiver architectures are discussed focusing on the feasibility of fully integrated implementation for multi-standard solution.

1. Receiver Design Basics

Sensitivity and selectivity are the two key specifications of a receiver. Sensitivity presents the ability of a receiver to demodulate a small desired signal in the presence of thermal noise (generated from both the environment and the circuits themselves) at an acceptable bit error rate (BER) while selectivity is a measure of the ability of the receiver to demodulate a small desired signal in the presence of much stronger interferers (blockers) in adjacent frequency bands at the acceptable BER. Sensitivity can be expressed by the noise figure of the receiver while the selectivity can be derived by the dynamic range or linearity of building blocks in the receiver. Limited dynamic range, or circuit nonlinearity, of the receiver may introduce more interferers internally and thus influences the system selectivity.

1.1. Sensitivity and Noise Figure

A receiver's sensitivity is defined as the minimum signal level that the system can detect with acceptable signal-to-noise ratio, which is determined by the receiver's demodulation scheme. There are no universal standards for its measurement for different applications, such as AM receiver, FM receiver and digital communication receiver [30] because its measure depends upon specific signal characteristics. However, we can use noise figure (NF) as a measure of sensitivity. NF compares the total receiver noise with noise that would be present if the receiver generated no noise. It is also defined equivalently as the signal-to-noise ratio (S/N) of the receiver output to the S/N of the signal source. The relation between NF and sensitivity, $P_{in,min}$, is expressed as:

$$P_{in,min} = -174dBm/Hz + NF + 10logB + SNR_{min} \tag{2.1}$$

where B is the receiver's ultimate channel bandwidth expressed in hertz, $P_{in,min}$ is the minimum input level that achieves SNR_{min} at the output. The term -174dBm/Hz is the thermal noise power density of the 50Ω source resistance. Since SNR_{min} is determined by the modulation and demodulation scheme of the system, $P_{in,min}$ is only determined by NF. In other words, NF and sensitivity are two different terms but represent the same receiver's ability to deal with small signal input.

While sensitivity can be measured only after the whole receiver is completed, NF is meaningful for every building block in the receiver. Furthermore, Friis equation allows us to characterize NF of cascaded stages, which is given by:

$$NF_{tot} = 1 + (NF_1 - 1) + \frac{NF_2 - 1}{A_1} + ... + \frac{NF_m - 1}{A_1 ... A_{m-1}} \tag{2.2}$$

where NF_{tot} is the overall noise figure referred to the input port, NF_m is the $m-th$ stage's noise figure and A_m is the $m-th$ stage's power gain. The equation indicates that the noise contributed by each stage decreases as the gain preceding the stage increases, implying that the first few stages in a cascade are the most critical. Conversely, if a stage exhibits attenuation (loss), the noise figure of the following circuit is "amplified" when referred to the input of that stage. Specifically, the noise figure of a passive component, such as a passive filter, is equal to its loss. For example, if an RF band-select filter has 3dB attenuation, its noise figure is 3dB.

1.2. Nonlinearity and Intermodulation

Active devices, such as bipolar transistors and MOSFET transistors, are inherently nonlinear though they may be approximated as linear devices in a small operating region. When the input signal is large enough or the internal signal of the devices becomes large such that the operating region is out of the linear region, the nonlinearity comes into play. Nonlinearity is not always harmful. Mixers, oscillators, comparators, limiters are all nonlinear circuits and are necessary building blocks of a receiver. But in amplifiers, active filters, whose functions are to linearly amplify or attenuate the signal, nonlinearity is destructive.

The nonlinearity of a circuit can be expressed as a power series:

$$y(t) = k_1 f(t) + k_2 f^2(t) + k_3 f^3(t) + k_4 f^4(t) + k_5 f^5(t) + ... \tag{2.3}$$

where $f(t)$ is the input, k_i is the nonlinear coefficient and $y(t)$ is the output. If the circuit works in weak nonlinear region, the nonlinearity can be simplified as a third-order equation:

$$y(t) = k_1 f(t) + k_2 f^2(t) + k_3 f^3(t) \tag{2.4}$$

In practice, the nonlinearity of a circuit is often characterized in frequency domain by applying sinusoid signals. Let the input be the addition of m sinusoid tones:

$$f(X) = \sum_{n=1}^{m} A_n \cos X_n = \sum_{n=1}^{m} Re(A_n e^{jX_n}) \tag{2.5}$$

where $A_n = a_n A$ is the signal amplitude and $X_n = 2\pi f_n t + \Phi_n$ is the instant phase. Then the fundamental output of equation 2.4 is:

$$y_{fund} = \sum_{n=1}^{m} Re\{[k_1 + \frac{3}{4} k_3 (A_n{}^2 + 2 \sum_{p=1,p\neq n}^{m} A_p{}^2)] \times A_n e^{jX_n}\} \tag{2.6}$$

The gain normalized to k_1 is:

$$G = 1 + \frac{3}{4}\frac{k_3}{k_1}(A_n{}^2 + 2 \sum_{p=1,p\neq n}^{m} A_p{}^2) \tag{2.7}$$

Note that the gain G is a function of all input signal's amplitudes. Depending on the sign of k_3, it can be lower or higher than 1. In real circuits, normally k_3 is negative. As the input signal is larger, the gain G is smaller, or compressed. All third-order intermodulation products can be expressed as:

$$y_{imd} = \frac{3}{4} \sum_{n=1,p=1,p\neq n}^{m} Re[k_3 A_n{}^2 A_p e^{j(2X_n \pm X_p)}] \tag{2.8}$$

These harmonic and intermodulated products either deteriorate the signal or limit the circuits' dynamic range. In the transceiver front-end, it is very important to investigate the nonlinearity in both low noise amplifier of the receiver part and the power amplifier of the transmitter. Frequency planning should always be carefully designed to avoid having these intermodulation product lie in the signal band.

1.3. Selectivity

Selectivity is the property of a receiver that allows it to separate a signal or signals at one frequency from those at all other frequencies. At least two characteristics must be considered simultaneously in establishing the required selectivity of a receiver. The selective circuits must be sufficiently sharp to suppress the interference from adjacent channels and spurious responses. On the other hand, they must be broad enough to pass the highest sideband frequencies with acceptable distortion in amplitude and phase.

The overall receiver selectivity is divided among RF, IF, baseband and digital selective elements. Though an amplifier and mixer can have some selectivity, in the receiver design they are treated as broadband components. The selectivity is completed by all kinds of filters. Based on the specific filtering function, they can also be named as: band filter, channel filter, decimation filter etc. Along the receiver chain, there is high probability of nonlinearities in the active components, which generates overload, modulation distortion, spurious signals and spurious responses. Thus the filters not only need to filter out the interference in the original received signal, but also need to filter out any spurious and distortion that are generated by the receiver itself. So different architectures and different frequency plans bring different filtering problems. Careful choice of receiver architecture and frequency plan can greatly relax the selectivity realization.

1.4. Dynamic Range

Dynamic range (DR) is generally defined as the ratio of the maximum input level that the circuit can tolerate to the minimum input level at which the circuit provides a reasonable signal quality. This definition is quantified in different applications differently [61]. In analog circuits such as op amps and analog-to-digital converters the dynamic range is defined using both signal-to-noise ratio (SNR) and spurious-free dynamic range (SFDR). Both these specifications are defined with the input signal at -3dB "full-scale". The ratio between the output signal power and the power of the largest spurious tone is SFDR while the ratio between output signal power and the total inband noise is SNR. In RF design, the lower end is defined by the sensitivity. The upper end of the dynamic range is defined as the maximum input level in a two-tone for which the third-order IM products do not exceed the noise floor. The concept of IIP3, input third-order intercept point, is used to quantify the intermodulation behavior. As shown in equation 2.8, the intermodulation product has a cubic relation with the input signal's amplitude. It increases faster than the output signal when the input signal is increasing. If we extrapolate the small signal nonlinear behavior based on equation 2.4, the intermodulation product in the two-tone test will be equal to the output signal. The input signal level at this point is called IIP3.

Expressing all of the quantities in dBm, an RF amplifier (LNA or PA)'s IIP3 can be calculated based on the weak nonlinearity assumption of equation 2.4:

$$P_{IIP3} = P_{in} + \frac{P_{out} - P_{IM,out}}{2} \tag{2.9}$$

where $P_{IM,out}$ denotes the power of IM_3 components at the output. Since $P_{out} = P_{in} + G$ and $P_{IM,out} = P_{IM,in} + G$, where G is the circuit's power gain in dB and $P_{IM,in}$ is the input-referred level of the IM_3 products, we have:

$$P_{IIP3} = P_{in} + \frac{P_{in} - P_{IM,in}}{2} = \frac{3P_{in} - P_{IM,in}}{2} \tag{2.10}$$

and hence:

$$P_{in} = \frac{2P_{IIP3} + P_{IM,in}}{3} \tag{2.11}$$

The input level for which the IM products become equal to the noise floor is thus given by:

$$P_{in,max} = \frac{2P_{IIP3} + F}{3} \tag{2.12}$$

where $F = -174dBm + NF + 10\log B$. The SFDR is the difference (in dB) between $P_{in,max}$ and $P_{in,min}$:

$$SFDR = \frac{2P_{IIP3} + F}{3} - (F + SNR_{min}) = \frac{2}{3}(P_{IIP3} - F) - SNR_{min} \tag{2.13}$$

The spurious-free dynamic range represents the maximum relative level of interferers that a receiver can tolerate while producing an acceptable signal quality from a small input level.

For a cascaded network, as shown in Figure 2.1, the total input referred $P_{IIP3cas}$ can be estimated by two methods. The first and simplest approach is to reflect each of the individual intercept points to the input and find the minimum term and approximate this as the intermodulation intercept point for the cascaded chain, which can be expressed as:

$$P_{IIP3cas} = min(P_{IIP3,1}, \frac{P_{IIP3,2}}{G_1}, \frac{P_{IIP3,3}}{G_1 G_2}) \tag{2.14}$$

It works well when trying to predict the intermodulation performance of a number of cascaded blocks when there is a "weakest link" in the chain such that its output intercept point dominates in the chain. However when the individual IP3 contributes somewhat equally to the overall chains' linearity performance then equation 2.14 is not a good approximation. The second approach attempts to take into account the interaction of the intercept points between the cascaded blocks in the chain. In this approach, the assumption is that the distortion contributions from the different blocks are uncorrelated, thus their distortion products are independent from block to block. The total input referred IP3 can then be calculated as:

$$P_{IIP3cas} = \frac{1}{\sqrt{1/P_{IIP3,1} + G_1/P_{IIP3,2} + G_1G_2/P_{IIP3,3}}} \tag{2.15}$$

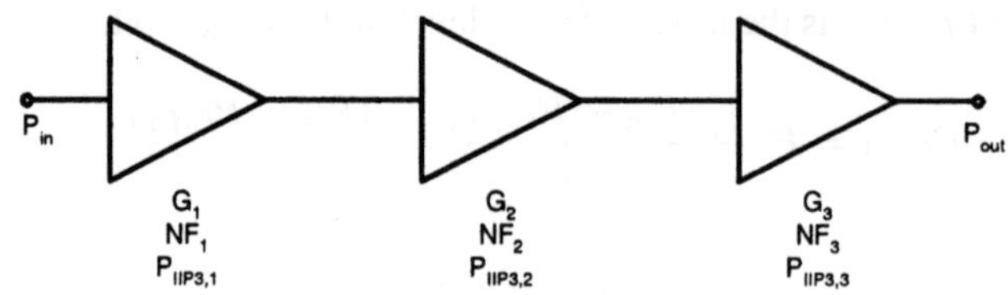

Figure 2.1. Cascaded Amplifier Blocks

2. Receiver Architecture Overview

Our study of the basic specifications of a radio receiver in section 1 treated the receiver as a black box. In this section, we discuss the receiver at the architecture level. Complexity, cost, power dissipation and integration level have been the primary criteria in selecting transceiver architecture. Furthermore, multi-standard capability is another important factor. As IC technologies evolve, the relative importance of these criteria changes, allowing approaches that once seemed impractical to return as plausible solutions.

2.1. Superheterodyne receiver

The superheterodyne receiver, shown in Figure 2.2 remains the architecture of choice for the vast majority of RF applications in the world today [55] [70] [85] [46] [77] [53]. Its constant popularity is due to its ability to reproducibly pick out narrow-bandwidth high-frequency signals from the surrounding background clutter of interferences outside the frequency range of interest. The RF signal is applied to a low-noise amplifier (LNA) and subsequently an image-rejection (IR) filter, which aims to filter out interferers at the signal's image frequency relative to the local oscillator frequency. The combination of extremely high performance and low power requirement result in the LNA being one of the most significant power drains in the system. The IR filter is typically implemented with a physically large surface acoustic wave (SAW) filter to provide sharp (high Q) filter function. In addition to their size, these filters have extremely unforgiving sensitivities to variations in source and load impedance, ground loops and so on. The resulted signal from the IR filter is mixed with the output of a local oscillator (LO), thus producing the intermediate-frequency (IF) signal. The IF filter suppresses out-of-channel interferers, performing channel selection.

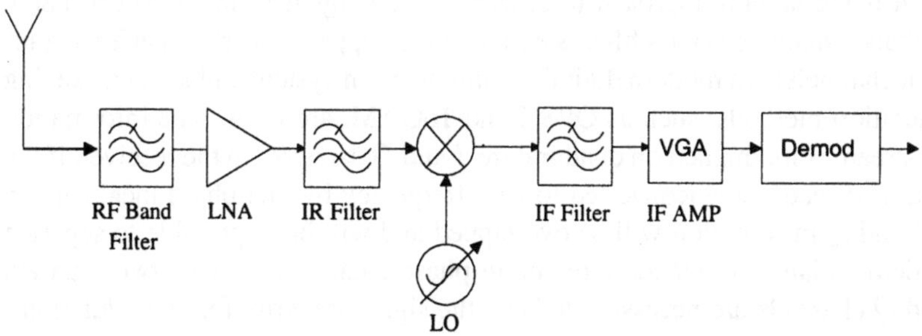

Figure 2.2. Superheterodyne Receiver Architecture

The principle issue in superheterodyne receiver is the tradeoff between IR and adjacent channel suppression. For given filter quality factors (Qs) and losses, if the IF is high, the image is greatly attenuated whereas nearby interferers remain at significant levels. Conversely, if the IF is low, the image corrupts the downconverted signal but the interferes are suppressed. To solve this problem, the concept of heterodyning can be extended to multiple downconverstions, each followed by filtering and amplification, which are called "dual-IF" and "triple-IF". However multiple downconversion brings some other problems. Each mixer generates many spurious or intermodulation components whose frequencies can be located near the RF, IF or baseband signal and LO signal. Some of these components may fall in the desired channel by more nonlinearity of the following circuits, degrading the signal quality. Thus, the "frequency planning" of the receiver plays a key role in the overall performance that can be achieved.

Another important drawback of heterodyne receivers is that the LNA must drive a 50Ω load because the IR filter is placed off-chip. Any deviation from 50Ω match will deteriorate the filtering transfer function. Despite the complexity and the need for a large number of external components, heterodyning is still viewed as the most reliable receiver architecture. However, in terms of frequency plan, the superheterodyne architecture is not suitable for single-chip multi-standard application. To deal with the different signals from 200KHz for GSM to 5MHz for WCDMA, the use of multiple off-chip image rejection filters and channel selection filters are inevitable for different standards, which eliminates the possibility of a high-integration and low cost handset receiver.

2.2. Zero-IF Receiver

To completely eliminate the image problem of superheterodyne receivers, the IF is set to zero frequency. This receiver architecture is called "Zero-IF" receiver. It is also called direct conversion receiver (DCR) or homodyne receiver. A diagram of a typical direct conversion receiver, is shown in Figure 2.3. It is the natural approach to downconvert a signal from RF to baseband. Baseband analog low-pass filter is employed to suppress nearby interferers (adjacent channels). In modern digital communication systems, phase shift keying modulation methods, such as QPSK and 16QAM, are used. The information data stream is modulated on quadurature I and Q carriers. In the receiver, if the signal is directly downconverted to zero frequency by one-phase local carrier, the I and Q information will be overlapped and will be impossible to separate and demodulate. So quadurature and in-phase local carriers with two separate I and Q channels are necessary to keep the signal integrity for demodulation.

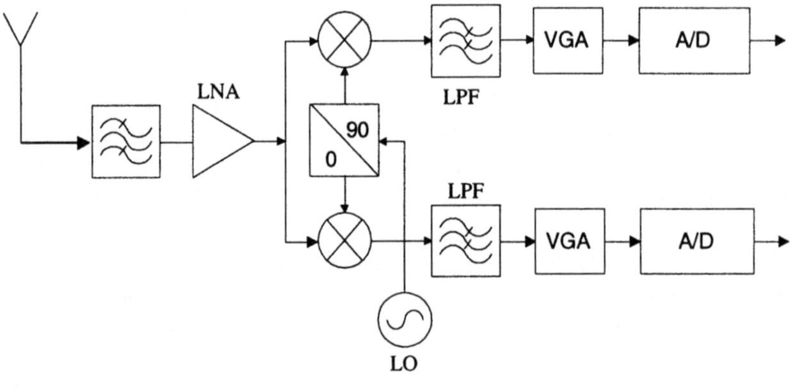

Figure 2.3. Direct Conversion Receiver Architecture Diagram

Direct conversion has several advantages over heterodyning. First the problem of image is circumvented because the IF is zero. Second, the LNA need not drive a 50Ω load because no image rejection filter is required. Third, the IF SAW filter and subsequent stages are replaced with low-pass filters and baseband amplifiers that are amenable to monolithic integration with very low power consumption.

In DCR, the LO frequency and the baseband low-pass filter can be made programmable to satisfy multiple communication standards. While the DCR approach is suitable for full integration and can be easily made multi-standard capable, several system level problems need to be overcome and actually they are all being actively pursued by a variety of worldwide research groups [65] [44] [75] [76] [86]. In general, these problems are: DC offset, even-order distortion, 1/f noise and LO leakage and self-mixing. All these offsets and low frequency noise should be removed either by analog feedback circuitry or adaptive digital cancellation. To illustrate the design challenges, a GSM receiver typically has a sensitivity of some -105dBm and an LO leakage as low as -85dBm will cause the DC level to be 10 times the signal amplitude at the range limit [50]. In Chapter 3, the detailed system analysis and frequency plan using DCR for multi-standard terminals are presented.

2.3. Low-IF Receiver

The low IF receiver shown in Figure 2.4 converts the received RF signal to a low intermediate frequency whereby an on-chip bandpass filter can be used to perform channel selection [12]. Though this architecture eliminates the problems of DC offset and 1/f noise associated with DCRs while maintaining the same level of integration, it reintroduces the image rejection requirement. This time because of the rather low IF frequency image rejection is difficult to implement. Image rejection mixer techniques are required instead of an image rejection filter, which is almost impossible to implement either on-chip or off-chip in this situation. In practical image rejection mixers, the amplitude and phase mismatches between I and Q channels are the ultimate limitation of the image rejection ratio. These mismatches come from the two-phase LO signals, I and Q mixers and the addition function. For typical matching in integrated circuits, image suppression falls in the range of 30 to 40dB while in high performance cellular phone applications, the suppression must be around 60 to 70dB [61]. A bandpass filter and/or a bandpass $\Delta\Sigma$ modulator are used to perform A/D conversion. Since given the same specifications for channel filtering, bandpass filter or modulator needs twice the number of poles and zeros compared to its lowpass counterpart and the power consumption to achieve a wide bandwidth standard, such as WCDMA, becomes prohibitive. Note that in Figure 2.4, when the signal bandwidth is comparable to the IF frequency, the bandpass A/D converters can be replaced by lowpass (Nyquist) A/D converters.

For specific modulation schemes, the low-IF receiver architecture is better suited and can be made simpler. In Bluetooth [72], which is a frequency-hopped GMSK/GFSK modulated spread spectrum system, the low-IF receiver can be implemented by a traditional FM analog demodulator, as shown in Figure 2.5. In this method, only 1-bit A/D converter is needed, which actually is a comparator. However external LC tank, or on-chip active inductor/gyrator, is required to realize the delay block in the FM demodulator. To make it fully-integrated, a higher resolution A/D converter can be used, shown in Figure 2.6. Then the channel filtering and demodulation are realized in the digital domain. This also allows more efficient solutions when migrating to deep sub-micron CMOS process, taking full advantage of the technology scale-down with digital design.

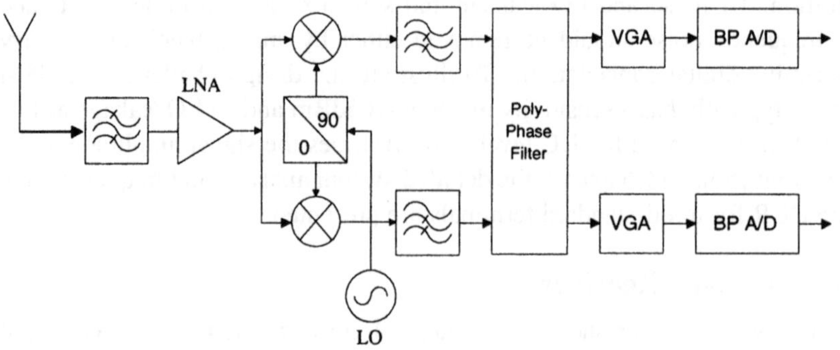

Figure 2.4. Low-IF Receiver Architecture with I and Q Channels

Note that the architecture no longer requires I and Q channels after the image rejection processing. This is the preferred fully-integrated receiver architecture for FM modulated signal. So it can be expanded to a multi-standard receiver for LAN/PAN, which use frequency hopping and GFSK/GMSK modulation schemes, such as IEEE 802.11 FHSS, HIPERLAN Type 1, HomeRF1.0 and HomeRF2.0 applications. Note that all the standards operate at 2.4GHz except HIPERLAN Type 1, for which we need a separate LNA and a multi-band LO in the RF part while the baseband is still the same.

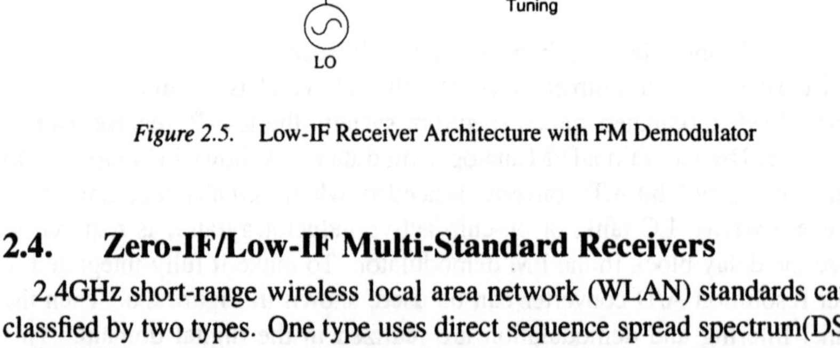

Figure 2.5. Low-IF Receiver Architecture with FM Demodulator

2.4. Zero-IF/Low-IF Multi-Standard Receivers

2.4GHz short-range wireless local area network (WLAN) standards can be classfied by two types. One type uses direct sequence spread spectrum(DSSS) and QPSK modulation techniques, including the high data rate versions of 802.11 DSSS .The other uses frequency hopping spread spectrum(FHSS) and

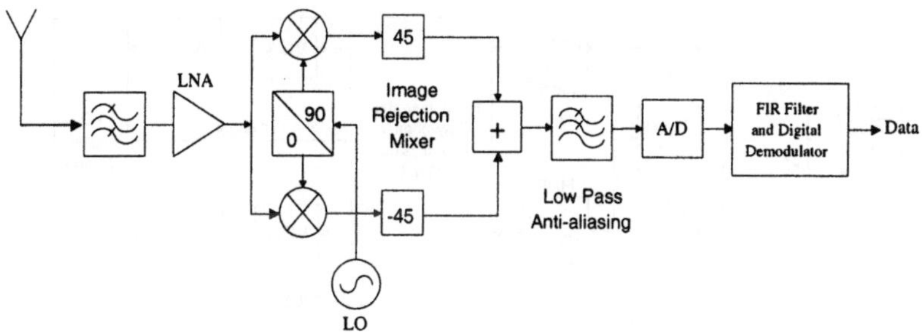

Figure 2.6. Low-IF Receiver Architecture with Digital Demodulator

GFSK modulation techniques, including 802.11 FHSS, Bluetooth and HomeRF v1.0. From the discussions in previous sections,one may realize that a multi-standard receiver architecture can be developed by a reconfigurable zero-IF/low-IF multi-standard architecture [78]. For example,combining the zero-IF receiver shown in Figure 2.3 with the low-IF receiver shown in Figure 2.6, the resulting zero-IF/low-IF ISM band multi-standard receiver, shown in Figure 2.7, can be reconfigured from the digital baseband and optimized to be suitable for both types of WLAN standards. Since all the standards use the same 2.4GHz ISM band, the front-end radio circuits can all be shared. For low-IF operation, which includes 802.11 FHSS, Bluetooth and HomeRF, the poly-phase filter is enabled by the control signal from the digital baseband and thus performs image rejection.The LO frequency is also switched, to generate the required low-IF signal with the mixer .Only one VGA and ADC channel is used.Channel selection is performed digitally for both types. In the digital baseband , programmable FIR filters and demodulation algorithms are implemented for different standards. In this receiver, except for the analog filter part, most of the analog circuits are shared, this leads to a low power, cost-effective multi-standard ISM band multi-standard receiver solution.

2.5. Wideband IF Double-Conversion Receiver

An alternative architecture well suited for integration of the entire receiver is wideband IF with double conversion [63]. Shown in Figure 2.8, this receiver takes all of the potential channels and downconverts them from RF to IF using a mixer with a fixed frequency local oscillator $LO1$. A simple low-pass filter is used at first IF to remove any upconverted frequency components, allowing all channels to pass to the second stage of mixers. All of the channels at first IF are then downconverted again to baseband using a tunable, channel-select frequency synthesizer $LO2$. Adjacent channel energy is then removed with a baseband filtering network where variable gain may be provided. This ap-

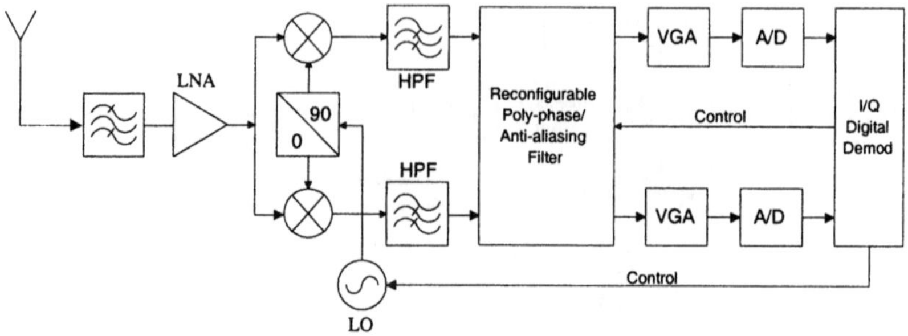

Figure 2.7. Zero-IF/Low-IF Multi-standard Receiver

proach is similar to a superheterodyne receiver architecture in that the frequency translation is accomplished in multiple steps. However, unlike a conventional superheterodyne receiver, the first local oscillator frequency translates all of the receive channels, maintaining a large bandwidth signal at first IF. The channel selection is then realized with the lower frequency tunable second LO. Same as in the case of direct conversion, channel filtering can be performed at baseband, where digitally-programmable filter implementations can potentially enable more multi-standard-capable receiver features. This is essentially a dual downconversion heterodyne receiver, where the first downconversion employs an image reject mixer, and the second follows the homodyne approach.

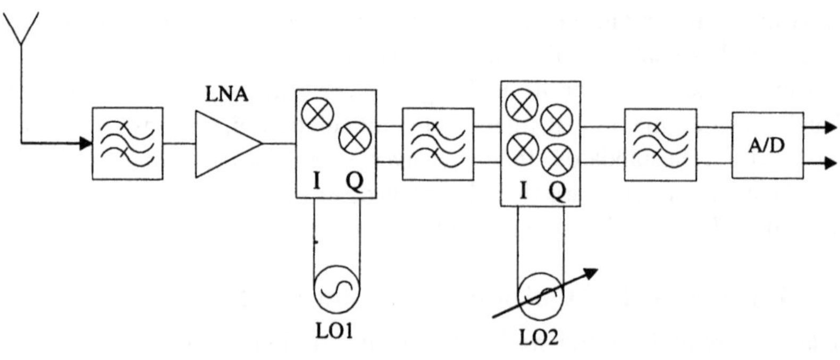

Figure 2.8. Wide-band Double IF Receiver Architecture Diagram

The advantages of this architecture are that it eases the generation of the first local oscillator, which is now at a fixed frequency and that reradiation of the local oscillator back to the antenna is not a problem, as it would be if a direct downconversion technique were employed. For multi-standard and multi-band applications, the first local oscillator can be tuned for different bands. This can be realized by switching among fixed value components, such as inductors or varactors, in the frequency synthesizer. So basically, the first local oscillator can be designed to have very good phase noise performance since the divider ratio can be set to be very small and thus greatly reduces the dividers' contribution to overall phase noise. This approach is highly desirable for monolithic integration, but suffers from the use of six high performance mixers to perform the complete downconversion, raising the DC power dissipation considerably. It also suffers from the same problem of existing homodyne receivers in terms of sensitivity to DC offsets, 1/f noise and second-order distortion since the second downconversion actually is a DCR. Image rejection problem is also limited by on-chip matching of building blocks as in low-IF receivers. From the frequency plan point of view, the nonlinearity of the six front-end mixers tends to generate a lot of intermodulation products that can be easily located in the signal bandwidth. So from the performance, cost and power consumption perspective, DCR is more suitable for multi-standard receivers.

2.6. Digital-IF Receiver

In the dual-IF heterodyne architecture, the second set of mixing and filtering, in which the signal is already downconverted to lower frequency, can be performed more efficiently in the digital domain. Shown in Figure 2.9 is an example where the first IF signal is digitized. In the digital downconversion part, the digitized signal is mixed with the quadurature phases of a digital sinusoid, and low-pass filtered to yield the quadurature baseband signals. This approach is called digital-IF architecture. This architecture is widely used in the state-of-the-art basestation receivers. It is also a strong candidate for future software radio receivers. The *I* and *Q* mismatch can be completely eliminated with digital processing.

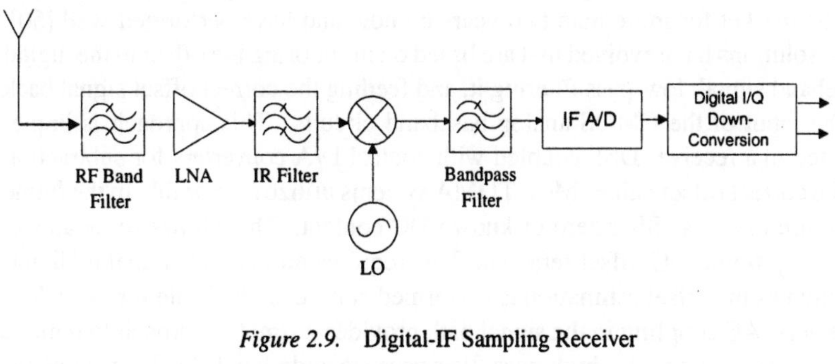

Figure 2.9. Digital-IF Sampling Receiver

The major issue in this approach is the performance required of the A/D converter. Since the first IF bandpass filter is a band filter, all the adjacent interferers or blockers are directly added at the A/D converter's input. Thus the

A/D Converter's dynamic range must be wide enough to accommodate both the high power level blockers and the very low power level signal while leaving the channel filtering to digital signal processing. On the other hand, to relax the image problem of the first downconversion, the IF frequency cannot be set very low, the normal value is around 50MHz to 200MHz. Note that the signal band is around 20MHz to 80MHz wide, for example 80MHz in the DCS1800 standard, 60MHz in the FDD WCDMA standard and 20MHz in the TDD WCDMA standard, which is comparable to the sampling frequency of the A/D converter, thus practically the A/D converter is realized by a lowpass converter rather than a bandpass converter, such as a bandpass $\Delta\Sigma$ modulator A/D converter. The A/D Converter's sampling frequency should be larger than 100MHz to 400MHz according to the Nyquist criteria. Such a high-resolution high-speed ADC is very challenging to implement. To relax the requirement, the sampling frequency can be reduced to the range of 50MHz to 100MHz by using sub-sampling technique. Even if this A/D Converter is available, the high power consumption that is associated with it makes it unsuitable for battery-operated mobile handsets. However it is a perfect solution for baseband software radio receivers if the required A/D converter performance can be obtained.

3. Summary

Among the five receiver architectures, zero-IF is the most promising candidate for multi-standard terminals as discussed above [21]. The DC offset problem has hindered its widespread adoption in modern high performance receivers. However recent extensive research has shown high potential of using it as the dominant low power, high integration receiver for future mobile communication terminals [67][68][84][16][66][58][28][45][32]. The zero-IF receivers for handsets, e.g. Ericsson's dual-band homodynes T28, have been on the market for more than two years by now and have performed well [50]. New solutions have evolved that are based on monitoring the offset in the digital baseband signal, low-pass filtering it, and feeding the correct offset signal back to the input of the CMOS analog baseband circuits. This approach is implemented in a receiver DSP coupled with control D/A converters for subtraction of the correct offset value. Most TDMA systems utilize a preamble in the frame structure that has either zero or known DC content. This allows for adaptive, frame-by-frame DC offset removal. This requires an integrated analog/digital solution as the offset estimation is performed at baseband. While for WCDMA receivers, AC coupling in the signal path provides a simple approach to remove the dc component with a high-pass filter because wide-band DS-SS systems are less sensitive to the DC notch. The spreading operation spreads each bit over pseudorandom sequence, which means that the loss of one information bit is an average over a period rather than a failure on a single transmission in the constellation. The averaging works better when the processing gain, defined as the ratio of chip rate to date rate, increases.

Besides the DC offset problem, the baseband processing and A/D converter in zero-IF receivers need higher dynamic range since there is no effective selective filtering functions in the LNA and Mixer. $\Delta\Sigma$ modulator A/D converter is an ideal candidate for this application, which will be discussed in detail in later chapters.

Chapter 3

MULTI-STANDARD RECEIVER ANALYSIS

The air-interface of a multi-standard receiver can be described by Figure 3.1. The RF input covers a wide range of RF frequency bands. Each standard has its own RF band and channel spacing. In this chapter, the WCDMA/DCS1800/DECT multi-standard receiver is analyzed. The receiver functions exactly as any single standard receiver under a standard-control signal. The receiver specifications are first derived based on the system requirement. Then the receiver specification is distributed to building blocks and thus block level specifications are obtained for all blocks from the LNA to the A/D converter.

Figure 3.1. Input and Output Spectrum of a Multi-Standard Receiver

1. WCDMA/GSM/DECT Multi-Standard Architecture

Receiver ICs for certain standards using zero-IF architecture, such as GSM, DBS, WCDMA, DECT, WLAN, are reported [67][68][84][16][66][58][28][45]

[32] in the recent literature. It is observed that zero-IF architecture, which has been proven for dual-mode GSM handsets [50], is receiving considerable attention, compared to other architecture, for future single-chip multi-standard receivers.

The air interfaces of the target multi-standard receiver are listed in Table 3.1. Note that TDD WCDMA is discussed here. Based on our detailed analysis of receiver architectures and the three standards, a zero-IF multi-standard receiver architecture is proposed [40]. Figure 3.2 depicts the proposed zero-IF radio architecture for a TDD UMTS (WCDMA/GSM) with DECT-supportive terminal. Because of the TDD operating mode the terminal does not require simultaneous transmission and reception, thus a TDD switch can be used at the front-end instead of a duplexer. For a FDD UMTS terminal, the TDD switch should be replaced by a duplexer for WCDMA while the TDD switch is still needed for GSM and DECT.

Table 3.1. Radio Specifications of WCDMA/DECT/GSM

Parameters	GSM (DCS1800)	FDD WCDMA	TDD WCDMA	DECT
Frequency Band TX/RX (MHz)	1710-1785 1805-1885	1920-1980 2110-2170	1900-1920 2010-2025	1880-1897
Sensitivity(dBm) @ BER=10^{-3}	-100	-110 12.2Kbps	-107 12.2Kbps	-83
Maximum Power Level (dBm)	-15	-25	-25	-23
Channel Spacing	200KHz	5MHz	5MHz	1.728MHz
Input Noise(dBm)	-120.8	-107	-107	-112.3
Required C/N(dB)	9	7.2	7.2	10.3

Because of the different operating frequency bands, three sets of band filters and LNAs, which are basically narrow band components, are required to complete the band selection and low noise amplification. Several digital control signals from digital baseband processor control the power supply of the three LNAs so that only one required LNA is powered on and the other two are powered down. Two I/Q mixers are shared by all three standards because mixers are basically wideband components. In the proposed receiver, the baseband filter, variable gain amplifier (VGA) and A/D converter are all shared by the three standards. However, the baseband filter functions as a channel filter for WCDMA standard only. For DECT/GSM standards, it operates as an anti-aliasing filter. The VGA's (variable gain amplifier) dynamic range is calculated to satisfy all the standards. The A/D converter is realized by a programmable $\Delta\Sigma$ modulator. The resolution and signal bandwidth of this A/D converter are controlled to satisfy the three standards. The decimation filter following the analog modulator performs channel filtering for the DECT/GSM standards. Using this architecture, the three standards signals are processed by the same receiver. Except for off-chip band filters and on-chip LNAs, all building blocks are shared.

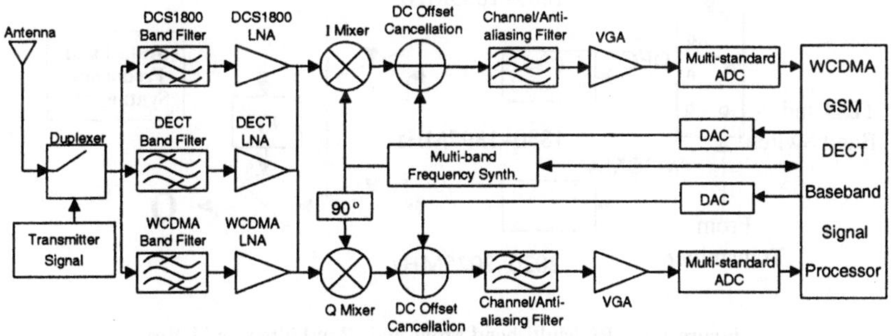

Figure 3.2. Proposed WCDMA/GSM/DECT Multi-Standard Receiver

Another approach to design the baseband chain and A/D converter is presented in [15][5][4], where digitally programmable analog filter and amplifiers are needed for channel selectivity thus achieving selectivity and amplification of all standards in the analog domain. All the standards share one fixed low resolution, 5 to 7 bit, A/D converter.

2. Front-end Circuit Design Issues

Here, we discuss design issues pertaining to the three main parts of the receiver, namely the RF, the analog baseband and the synthesizer.

2.1. RF circuits

In the multi-standard receiver, narrow band tuned LNAs are required to save power and achieve high gain in current sub-micron CMOS technologies. Since the frequency operating bands for GSM, DECT and WCDMA cover from 1805MHz to 2025MHz, three separate LNAs will be needed, one for each standard. A TDD and band selection switch is placed after the antenna. This switch is a single-pole-four-throw switch. The four throws are respectively WCDMA, DECT and the GSM receiver modes and for transmission. It is controlled by signals from the DSP controller chip, which is shown in Figure 3.3 as Standard-Switching Signal. This signal also controls the on-off status of the three LNAs so that only one LNA is turned on when the receiver is operating. Considering the trade-off between dynamic range and sensitivity, each LNA should have switchable high gain and low gain mode.

The I and Q mixers are shared by all standards. Though highly-linear passive mixers in CMOS are successful, the importance of having combined gain

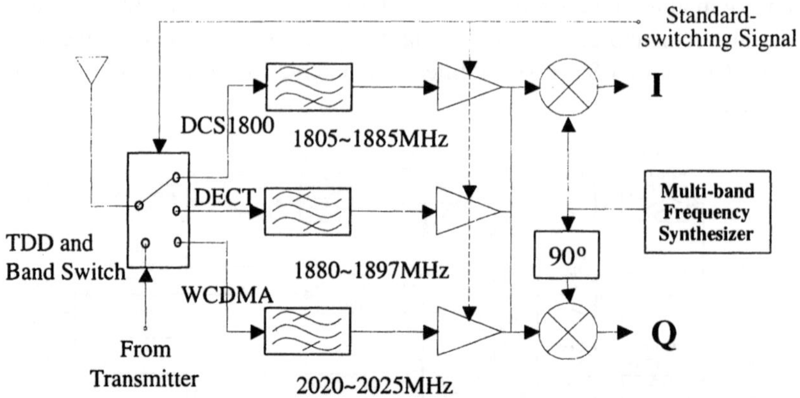

Figure 3.3. RF Multi-band Front-end: Band filters and LNAs

from the LNA and mixer, which should be sufficient to overcome flicker and thermal noise in the baseband stages, rules out lossy passive mixers in zero-IF receiver. The IIP_3 of a CMOS cascode LNA is quite high, typically greater than $-5dBm$ if noise and matching requirements are met. Thus, the linearity of the subsequent stages tends to be the limiting factor. In particular, the downconversion mixers must achieve high linearity and a reasonable noise figure. Single balanced mixer exhibits less input noise than that of double-balanced topology. But the double-balanced topology entails much less LO-IF feedthrough and suppresses the effect of additive noise in the LO input. CMOS mixers typically demand large LO swings so that the switching pairs do not remain on simultaneously for a considerable period of time. Increasing the width of the switching devices can lower the required swing, but at the cost of increasing their noise contribution and higher capacitance in the RF signal path. Thus, the choice of device dimensions and bias current plays a critical role in the performance.

The interface between the LNA and the mixer needs particular attention. In a heterodyne receiver, an external image-reject filter is ac-coupled to the LNA at its input and subsequently to the mixer at its output. These two interfaces are both 50Ω matched. In zero-IF architecture the impedance level and the coupling method at the interface between the LNA and the mixer are flexible. However, at the LNA's output node, either an on-chip inductor or off-chip inductor is necessary to tune out the capacitance at the interface node. Otherwise, it is impossible to get high gain at GHz range. A low-Q on-chip spiral inductor is preferred here because it can realize a broadband LNA gain without external tuning component. To eliminate even-order distortion from the LNA, a high pass network including an ac-coupling capacitor between the LNA and mixer is an easy and effective solution.

2.2. Baseband Circuits

The design of the receiver baseband circuits becomes progressively more difficult as the RF section incorporates fewer external components and hence performs a less portion of the overall signal processing task. Furthermore, noise-linearity-power trade-offs limit the amount of gain provided by the LNA and mixer circuits. After the signal is downconverted to the baseband, it must be filtered, amplified and digitized. At the interface node between the mixer and the first baseband stage, the signal is still quite small and the interferers may be quite large (e.g. 60dB above the signal level). Thus, both the noise and the nonlinearity of the first baseband building block are critical. Second, to avoid lowering the voltage gain of the mixer, it must exhibit a relatively high input impedance.

The baseband processing includes filtering and VGA. In this multi-standard receiver architecture the analog baseband filtering processing is designed to satisfy the channel filtering of the WCDMA standard. The channel filtering of the GSM and DECT standards are realized by A/D converter's decimation filter. There are two branches for I and Q channels. Figure 3.4 presents one possible architecture of the baseband processing [5]. The input differential signals come from the output of the mixer. Anti-aliasing filter is used to filter out the noise and blockers outside of the sampling frequency of the A/D converter. The combination of two Filter sections has enough attenuation of the blockers for the WCDMA standard. VGA1 and VGA2 realize the VGA function such that enough dynamic range is achieved. The interleaving of filtering and VGA helps to meet the required IIP3 specification of the whole receiver.

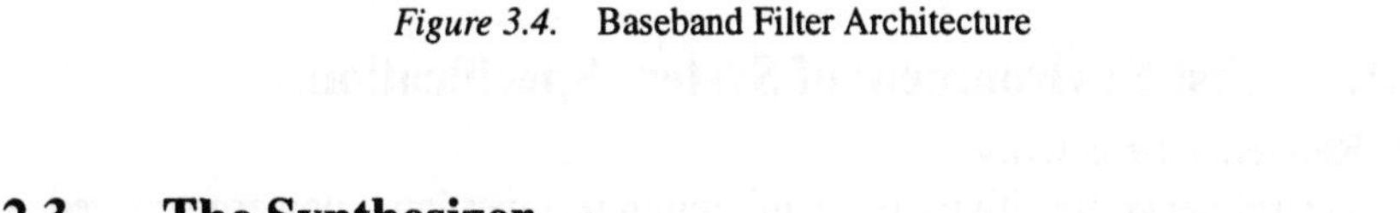

Figure 3.4. Baseband Filter Architecture

2.3. The Synthesizer

In Zero-IF receivers, the local oscillator frequency is always set to the center frequency of the desired signal. For the multi-standard zero-IF receiver, the LO is also switched among the three frequency bands. Since the total frequency band of the receiver is from 1805MHz to 2025MHz, which is not too large for the tuning range of one single voltage-controlled oscillator (VCO), it can be implemented either by one VCO or several VCOs for high phase-noise performance. Considering the multiple operating bands and different channel bandwidths, the frequency synthesizer architecture needs to be a fractional PLL and an array of loop filters should be used to optimize the loop transfer function such that the optimized switching time, spurious response and phase noise can be obtained for each standard. If a single VCO is used, the phase noise performance of the VCO must be good through a wide range of offset frequency. One more easy method is to use multiple varactors to cover different RF bands. The phase noise performance requirement will be discussed in the following sections.

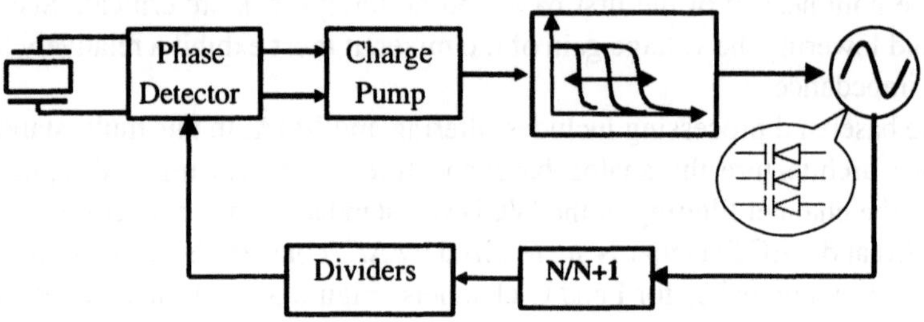

Figure 3.5. Local Oscillator Architecture

3. Radio System Analysis

In this section, the system requirement of the radio physical layer for the multi-standard receiver is described. This includes sensitivity, blocking characteristics and intermodulation characteristics. These test conditions given in the standard specification document [18][19][1][2] are closely related to the receiver's specifications, such as: noise figure, IIP3, dynamic range and filtering processing. The three standards, WCDMA, GSM(DCS1800) and DECT are then analyzed one by one. In the end, the multi-standard receiver's specification is concluded.

3.1. Test Environment of System Specification

- Reference Sensitivity
  The reference sensitivity is the minimum receiver input power measured at the antenna port at which the Bit Error Rate (BER) does not exceed a specific value. The noise and sensitivity specifications of WCDMA/GSM/DECT standards are shown in Table 3.1. The required Noise Figure (NF) requirement of the receiver can be calculated by rearranging equation 2.1:

$$NF = P_{in,min} - SNR_{min} + 174dBm/Hz - 10logB \tag{3.1}$$

Note that for a WCDMA system, another factor, which is the processing gain (PG), should be included in the calculation. According to the WCDMA standard document, the sensitivity requirement is defined for the signal bit rate of 12.2kpbs while the chip rate is 3.84Mbps. Thus the processing gain, which is the ratio between the two rates, is 25dB. Because of the processing gain, the required NF is greatly relaxed.

- Blocking Requirement
  The blocking characteristic is a measure of the receiver's ability to receive a wanted signal at its assigned channel frequency in the presence of an unwanted interferer on frequencies other than those of the spurious response or the adjacent channels, without this unwanted input signal causing a degradation of the performance of the receiver beyond a specified limit. Blocking requirement mainly affects the dynamic range and filtering of the receiver building blocks.

- Intermodulation Requirement
  Third and higher order mixing of two interfering RF signals can produce an interfering signal in the band of the desired channel. Intermodulation response rejection is a measure of the capability of the receiver to receive a wanted signal on its assigned channel frequency in the presence of two or more interfering signals, which have a specific frequency relationship to the wanted signal. The intermodulation requirements for the three standards are shown in Table 3.2. The required IIP_3 can be obtained using the information in Table 3.1 and Table 3.2.

Table 3.2. WCDMA/GSM/DECT Intermodulation Requirement

	WCDMA	GSM	DECT
Desired Signal Level(dBm)	-107	-99	-80
Interferer Level(dBm)	-46	-49	-46

3.2. WCDMA System Analysis

WCDMA sensitivity is specified as the minimum input power measured at the antenna at which the BER does not exceed 10^{-3}. For this BER, the received signal power level should be less than or equal to $-117dBm/3.84MHz$. Also the signal is defined as having a data rate of 12.2kps and a spread chip rate of 3.84Mbps. So the processing gain (G_p) of the system can be calculated as:

$$G_p = 10 \times log(\frac{3.84Mcps}{12.2Kbps}) = 25dB \tag{3.2}$$

Based on the 3GPP Radio Access Networks Technical Specification Group [1] [2], the E_b/N_0 required for the baseband receiver for a BER of 10^{-3} is 5.2dB. Including a 2dB implementation margin, it would be 7.2dB. Then the receiver NF can be calculated using equation 3.1 with processing gain:

$$NF_{WCDMA} = -117dBm - 7.2dB - (-174dBm/Hz)$$
$$-10log(3.84 \times 10^6) + 25dB = 9.0dB \qquad (3.3)$$

Stringent dynamic range specifications place large signal handling requirement on the WCDMA receiver mode. In WCDMA, the BER cannot exceed 10^{-3} when an average input power of -25dBm/3.84MHz is applied at the antenna. Given a peak-to-average power ratio of about 7dB for the WCDMA signal, the peak power at the antenna could approach -18dBm/3.84MHz. As the level of the received signal approaches the receiver's -1dB compression power, the distortion increases and thus reducing the E_b/N_0 and increasing the BER. A low-gain mode of LNA is used to relax the requirement of the following building blocks, especially the receiver AGC range. When the LNA is in low-gain mode, the input signal level is less amplified and the linearity and current consumption requirement of the VGA can be reduced.

In WCDMA the adjacent channel selectivity (ACS) is defined as a measure of the receiver's ability to receive a WCDMA signal at the assigned channel frequency, in the presence of an adjacent modulated channel signal at 5MHz offset from the center frequency of the assigned channel. ACS is the ratio of the receive filter attenuation at the assigned channel frequency to the receiver filter attenuation at the adjacent channels. The power of the wanted signal is -93dBm/3.84MHz while the modulated interferer is -52dBm/3.84MHz. Together with the blocking characteristics test shown in Table 3.3, two specifications need to be derived. One is the selectivity of the filtering process and the other is the LO's phase noise performance.

Table 3.3. WCDMA Blocking Requirement

In-band Offset(MHz)	Blocking(dBm)	Out-of-band Offset(MHz)	Blocking (dBm)
$10 < \lvert f - f_o \rvert < 15$	-56	$2050 < f < 2095$ $2185 < f < 2230$	-44
$\lvert f - f_o \rvert > 15$	-44	$2025 < f < 2050$ $2230 < f < 2255$	-30
		$1 < f < 2025$ $2255 < f < 12750$	-15

First let us calculate the LO phase noise requirement. At the mixer output, the blocker signal reciprocal mixing with the LO phase noise creates an interferer within the desired signal band. If the phase noise is assumed to be flat across the band of interest, the power of this interferer is:

$$P_I = P_B(\Delta f)(dBm) + [PN(\Delta f) + 10log(BW)](dBm) \tag{3.4}$$

where P_I is the reciprocal mixing product, P_B is the corresponding blocker power at the mixer output, Δf is the offset frequency, BW is the bandwidth of the signal and PN is the phase noise spectrum density. In the WCDMA mode, the 5MHz adjacent channel signal power is $-52-(-93)=41dB$ higher than the desired signal, then the difference between P_I and the desired signal is $41dB + [PN(\Delta f) + 10log(BW)]$. To keep this interference 15dB lower than the signal, then the phase noise requirement is:

$$PN(\Delta f) = -41dB - 10log(BW) - 15dB \tag{3.5}$$

Here we have BW is 3.84MHz, then the phase noise at 5MHz offset should be less than -125dBc/Hz. To make it more general, the phase noise requirement can be expressed as:

$$PN(\Delta f)(dBc/Hz) = P_S(dBm) - P_B(\Delta f)(dBm) - 10log(BW) - SIR(dB) \tag{3.6}$$

where SIR is the expected signal to the reciprocal mixing interferer ratio. Using the same equation, we can get the phase noise at 10MHz and 15MHz offsets according to the blocking characteristics shown in Table 3.3. In the blocking characteristics test, the desired signal level is -107dBm/3.84MHz. So for the phase noise at 10MHz offset:

$$\begin{aligned} PN_{10MHz} &= -107(dBm) - (-56dBm) - 10log(3.84 \times 10_6) - 15 \\ &= -132dBc/Hz \end{aligned} \tag{3.7}$$

and for the phase noise at 15MHz offset:

$$\begin{aligned} PN_{15MHz} &= -107(dBm) - (-44dBm) - 10log(3.84 \times 10_6) - 15 \\ &= -144dBc/Hz \end{aligned} \tag{3.8}$$

3GPP ACS test already gives the filter attenuation as 33dB at 5MHz offset. The baseband filtering attenuation of the receiver at 10MHz and 15MHz offset can be easily calculated as:

$$\begin{aligned} A_{10MHz} &= P_{B,10MHz} - P_S + SIR \\ &= -56dBm - (-107dBm) + 15dB = 66dB \end{aligned} \tag{3.9}$$

$$A_{15MHz} = P_{B,15MHz} - P_S + SIR$$

$$= -44dBm - (-107dBm) + 15dB = 78dB \qquad (3.10)$$

According to the intermodulation characteristics test environment, two interfering signals at 10MHz and 20MHz offset frequency are both at -46dBm power level. The input signal is at -107dBm and the receiver's BER should be kept lower than 10^{-3}. As we mentioned earlier, this BER corresponds to 7.2dB C/I, i.e. carrier-to-interference ratio, at the output of the receiver. Though it is an intermodulation test, it also includes the effects of noise in the receiver channel. Therefore, the distortion (intermodulation) components plus the white noise in the receiver together should satisfy the C/I requirement. Actually the desired signal level is 3dB above the sensitive requirement. So what the receiver needs to do is to manage the ultimate intermodulation product as low as the white noise power level. Then the receiver's C/(I+N), where N stands for noise, is just enough for BER requirement considering that the white noise is uncorrelated with the intermodulation components. According to the noise figure calculation, we know that the maximum receiver input referred noise floor is at $-117dBm - 7.2dB + 25dB = -99.8dBm$. So we also want all the intermodulation components from the receiver to remain less than -99.8dBm when referred to the receiver input. If the two intermodulating adjacent channels are applied to the receiver at -46dBm then we know that the IM3 component must be:

$$IM3 = -46dBm - (-99.8dBm) = 53.8dBc \qquad (3.11)$$

The input referred IP3 can be then calculated as:

$$IIP3 = -46dBm + \frac{IM3}{2}dBc = -19.1dBm \qquad (3.12)$$

3.3. DCS1800 System Analysis

The sensitivity requirement for DCS1800 is -100dBm with the required C/I of 9dB at the receiver output so that the BER can be kept lower than 10^{-3}. This reflects to a noise figure of 11.98dB.

$$NF_{DCS1800} = -100dBm - 9 + 174dBm/Hz - 10log(200\times10^3) = 11.98dB \qquad (3.13)$$

The blocking test for DCS1800 is performed by applying a GMSK modulated desired signal [20] 3dB above the required receiver reference sensitivity. Then a single unmodualted tone is applied to the receiver at discrete increments of 200KHz from the desired signal with a magnitude shown in Table 3.4. In our proposed receiver, the analog filter has no attenuation of the blockers. Thus a very high-resolution A/D converter is needed. The blocking characteristics here mainly influence the specification of the ADC. This will be discussed in the following sections. The phase noise specification of the LO can also be

Table 3.4. GSM(DCS1800) Blocking Requirement

In-band Offset(MHz)	Blocking(dBm)	Out-of-band Offset(MHz)	Blocking (dBm)
$0.6 < \|f - f_o\| < 1.4$	-43	$f < 1920$	-26
$1.6 < \|f - f_o\| < 2.8$	-33	$1920 < f < 1980$ $1705 < f < 1785$	-12
$\|f - f_o\| > 3.0$	-26	$f > 1980$ $f < 1785$	0

derived using equation 3.6.

$$PN_{600KHz} = -97 - (-43) - 10log(2 \times 10^5) - 15 = -122dBc/Hz \quad (3.14)$$

$$PN_{1.6MHz} = -97 - (-33) - 10log(2 \times 10^5) - 15 = -132dBc/Hz \quad (3.15)$$

$$PN_{3.0MHz} = -97 - (-26) - 10log(2 \times 10^5) - 15 = -139dBc/Hz \quad (3.16)$$

According to the definition of intermodulation test in DCS1800, the receiver must maintain 9dB C/I at the output of the receiver with the adjacent channel power of -49dBm. The maximum receiver input referred noise floor is at -109dBm and the desired signal is at -97dBm. The IM3 component must be:

$$IM3 = -49dBm - (-109dBm) = 60dBc \quad (3.17)$$

Therefore, the input referred IP3 is,

$$IP3 = -49dBm + \frac{IM3}{2}dBc = -19dBm \quad (3.18)$$

Which gives us a -19dBm input referred IP3 or better required of the receiver to be compliant with the DCS1800 standard.

3.4. DECT System Analysis

The sensitivity for DECT is -83dBm with required C/I of 10.3dB at the receiver output so that the BER can be kept lower than 10^{-3}. So the noise figure of DECT is:

$$NF_{DECT} = -83dBm - 10.3 + 174dBm/Hz - 10log(1.728 \times 10^6) = 18dB \quad (3.19)$$

DECT has considerably easier blocking requirements as compared to DCS1800. A set of test conditions is given for both inband and out-of-band blocking signals. For the inband blocking, the receiver must maintain 10^{-3} BER when a -73dBm desired signal is applied to the receiver and a single blocker is applied to the input of the receiver. The blocker is a GMSK modulated signal of power levels given in Table 3.5. The blocking requirements include an -83dBm Co-Channel blocker (The Co-Channel blocker is an interferring signal applied in the same band as the desired signal). All of the inband blocking tests are repeated for each of the adjacent channels. For the out-of-band blocking, a desired -73dBm input signal is applied to the receiver in channel 4, which is located at 4 times the channel spacing away from the desired signal. Then a single unmodulated blocker is applied in each of the bands with the signal strength indicated in Table 3.5. Same as the situation for DCS1800, the analog baseband filter though has some attenuation on the blockers, is not a channel filter. So the analog filter's attenuation characteristic is not important for the DECT mode. The blocking test will influence the ADC specification. This will be fully analyzed in the following sections. The phase noise requirement for DECT LO can be calculated, using equation 3.6.

$$PN_{2.2MHz} = -80 + 58 - 10log(1.728 \times 10^6) - 15 = -99dBc/Hz \quad (3.20)$$

$$PN_{3.9MHz} = -80 + 39 - 10log(1.728 \times 10^6) - 15 = -108dBc/Hz \quad (3.21)$$

$$PN_{5.6MHz} = -80 + 33 - 10log(1.728 \times 10^6) - 15 = -114dBc/Hz \quad (3.22)$$

Table 3.5. DECT Blocking Requirement

In-band Offset(MHz)	Blocking(dBm)	Out-of-band Offset(MHz)	Blocking (dBm)
$\|f - f_o\| < 2.2$	-58	$1950 < f < 2000$ $1780 < f < 1875$	-43
$2.2 < \|f - f_o\| < 3.9$	-39	$f > 2000$ $f < 1780$	-23
$3.9 < \|f - f_o\| < 5.6$	-33		
$5.6 < \|f - f_o\| < 9.0$	-33		

Note that the phase noise requirement is pretty relaxed compared with that of DCS1800. Similar to WCDMA and DCS1800, the DECT standard outlines a set of conditions to test the intermodulation performance of the receiver. A desired signal is at -80dBm. Two adjacent channel signals are applied with a -46dBm input power. Using the same procedure to calculate the IIP3 , we have an IIP3 for DECT of:

$$IM3 = -46dBm - (-94dBm) = 48dBc \quad (3.23)$$

$$IIP3 = -46dBm + \frac{IM3}{2}dBc = -22dBm \quad (3.24)$$

3.5. Baseband Filter and ADC dynamic range

To follow the receiver chain specification, in full details, is a complex task. Based on the proposed multi-standard WCDMA/GSM/DECT receiver architecture, the baseband filter order and ADC dynamic range can be determined first. In the receiver, the baseband filter operates as a channel filter for WCDMA standard while for DCS1800/DECT standards, it is an anti-aliasing filter. Because the anti-aliasing filter's specification is rather easy to realize, the filter specification is mainly determined by WCDMA blockers requirement and ADC's dynamic range. Actually the order of the filter and ADC's dynamic range can be traded with each other. A higher order filter leads to a low-resolution ADC while a lower order filter must be combined with a high-resolution (dynamic range) ADC to cope with the signal and the blockers.

First, the filter characteristic can be easily derived based on the WCDMA system analysis in the previous section. This is shown in Table 3.6 together with the ideal 6-th order and 5-th order butterworth filter specifications. In the table, the offset frequency means the offset from the center of the desired signal. It is easy to conclude that a 6-th order filter is required for the baseband filter.

Table 3.6. Baseband Filter Specification

Offset Frequency(MHz)	Attenuation Requirement(dB)	6th-order Attenuation(dB)	5th-order Attenuation(dB)
2.5	3	3	3
5.0	33	36	30
10.0	66	72	60
15.0	78	93	77

Actually the baseband filter has some attenuation for the DCS1800 and DECT's blockers, though the attenuation is not adequate. However here we do not consider the small amount of attenuation from the filter in DCS1800 and DECT operation because we assume that the baseband filter is in low power mode in DCS1800/DECT modes. Under the low power mode the filtering characteristic is fairly relaxed and thus we cannot expect to have sharp attenuation of the blockers. Then the dynamic range of the ADC for DCS1800 and DECT standards can be calculated only based on the blocker test specifications. For DCS1800 the signal is at the reference level -97dBm together with the highest blocker, which is -26dBm at 3MHz. If we set the ADC's quantization noise to be 15dB lower than the desired signal, the ADC's dynamic range is:

$$DR_{GSM} = -26dBm - (-97dBm) + 15dB = 86dB \tag{3.25}$$

For DECT standard, the signal power is -80dBm and the blocker level is -33dBm at 5.184MHz and 6.192MHz. Still we set the ADC's quantization noise be 15dB lower than the desired signal, the ADC's dynamic range for DECT is:

$$DR_{DECT} = -33dBm + 6dB - (-80dBm) + 15 = 68dB \quad (3.26)$$

In WCDMA ACS test, the signal power is -93dBm while the adjacent channel interferer is at -52dBm. Considering the channel filter attenuation on the adjacent channel to be 33dB, the adjacent channel interferer is still 8dB higher than the signal. Assume that the quantization noise in this mode is 20dB lower than the thermal noise, the ADC's dynamic range for WCDMA can be estimated as:

$$DR_{WCDMA} = -52dBm - 33 - (-93dBm) + 7.2 + 20 = 35.2dB \quad (3.27)$$

So the baseband filter and ADC's specification are set. The low-pass filter in the receiver is set to 6-th order and the ADC's dynamic range is 14bit, 11bit and 6bit for DCS1800, DECT and WCDMA respectively.

Up to now, we can summarize the system specifications for the three standards as shown in Table 3.7.

Table 3.7. WCDMA/DCS1800/DECT System Specification

	WCDMA	DCS1800	DECT
Noise Figure(dB)	9	12	18
IIP3(dBm)	-19	-19	-22
ADC Dynamic Range (dB)	36 6bit	86 14bit	68 11bit
Baseband filter Attenuation(dB)	33@5MHz 66@10MHz 78@15MHz	N/A	N/A
LO phase noise (dBc/Hz)	-132@10MHz -144@15MHz	-122@600KHz -132@1.6MHz -139@3.0MHz	-99@2.2MHz -108@3.9MHz -114@5.6MHz

3.6. Noise Figure, Gain and IIP3 Distribution

A TDD switch and a band-select filter are the two only off-chip components needed. The insertion loss of these components is set by checking the available components on the market [6] [54]. In our system, the total attenuation of these two components is set to be 4dB. Then the gain and noise figure of LNA and mixer are to be determined. Because we cannot afford a high noise figure deterioration from a passive Balun in the front-end, the three LNAs have a single ended input interface. The input must be matched to a standard 50Ω load. The gain and noise figure specifications were determined, and based on state-of-the-art published RF CMOS circuits performance. The results are listed in Table 3.8. Note that in the table, only high-gain values for LNAs are shown. These values are used when the received signal power is low. When the received signal at the antenna is high, the low-gain mode of LNAs is used, which is set to be 0dB. Though the gain is 0dB, the receiver can not just pass through the signal from the LNA input directly to the mixer since the LNAs' another important function is to provide enough reverse isolation to alleviate DC offset problem. The reverse isolation is even more important in the low gain mode.

Table 3.8. Building Blocks Specification Distribution

	WCDMA	DCS1800	DECT
Required NF	9dB	12dB	18dB
TDD Switch Insertion Loss(dB)	2dB	2dB	2dB
RF Band Filter Attenuation (dB)	2dB	2dB	2dB
Required Receiver IC NF	5dB	8dB	14dB
LNA NF	2dB	3dB	5dB
LNA Gain	18dB	16dB	12dB
IIP3	-10dBm	-10dBm	-10dBm
Mixer NF	12dB	12dB	12dB
Gain	8dB	8dB	8dB
IIP3	8dBm	8dBm	8dBm
Baseband NF	30dB	30dB	30dB
Gain	26-71dB	26-71dB	26-71dB
IIP3	16dBm	16dBm	16dBm

To adjust the signal level at an appropriate level when it hits the input port of the A/D converter, a variable gain amplifier(VGA) is required in the AGC loop. The dynamic range of the VGA is calculated by investigating the large-signal and small-signal operation. First, let us calculate the dynamic range for the WCDMA mode. When the signal is at the sensitivity level, which is -110dBm/3.84MHz or -93dBV, the whole receiver chain need offer 0dBV-(-93dBV)=93dB gain. Subtracting the gains of band-select filter (-2dB), TDD switch(-2dB), LNA(18dB) and mixer(8dB) in the receiver chain, the maximum VGA gain is 71dB. The maximum input power is -18dBm/3.84MHz or -33dBV, as pointed out earlier. At this condition, the LNA is set at the low-gain mode, which is 0dB. Thus the minimum gain from VGA is 33+2+2-0-8=26dB. So the VGA gain range is 26dB to 71dB, for a total of 45dB dynamic range. For

DECT and DCS1800, the required dynamic range of the VGA is smaller than that of WCDMA. So the VGA is set based on the WCDMA standard.

The IIP3 requirement for the three standards is about -19dBm as calculated in the previous section. However in the physical implementation of the handset, the amount of transmitter leakage into the receiver front-end desensitizes the LNAs if its linearity is insufficient. To prevent this desensitization, expressed in equation 2.7, the LNAs IIP3 or input P_{1dB} must be increased above those requirements set by the intermodulation test. Here we set IIP3 at -10dBm. According to equation 2.15, we evenly distribute the IIP3 specifications to mixer and analog low-pass filter. The result is that the IIP3 of the mixer is 8dBm and the IIP3 of the analog low-pass filter is 16dBm.

4. $\Delta\Sigma$ Modulators and Multi-Standard Receivers

The $\Delta\Sigma$ modulator ADC constitutes a major part of the proposed multi-standard receiver. As mentioned in the previous part, only WCDMA channel filtering is realized in the analog baseband processing. The filtering of the baseband circuits has almost no attenuation on the blockers of GSM and DECT. Thus for GSM and DECT standards, the ADC needs much higher dynamic range (resolution) compared with the WCDMA standard. A high performance programmable $\Delta\Sigma$ modulator ADC provides a good solution for meeting these requirements.

Conventional Nyquist architectures, are often difficult to implement very high dynamic range converters in fine-line very large scale integration technology. These difficulties arise because conventional methods need precise analog components in their filters and conversion circuits and because their circuits can be very vulnerable to noise and interference. The virtue of conventional methods is their use of a low sampling frequency, usually the Nyquist rate of the signal. Oversampling converters can use simple and relatively high-tolerance analog components to achieve high resolution, but they require fast and complex digital signal processing stages. These converters modulate the analog signal into a simple code, usually single-bit words, at a frequency much higher than the Nyquist rate. Thus the design of the modulator trades resolution in time for resolution in amplitude in such a way that imprecise analog circuits can be tolerated.

Oversampling $\Delta\Sigma$ Modulators have been very successfully used in A/D conversion applications over the last two decades [74]. Although the emphasis has been usually on low-speed, high-resolution applications, high-speed and high-resolution converters using oversampling $\Delta\Sigma$ Modulators have recently been reported [48] [52] [7]. Basically, a $\Delta\Sigma$ Modulator exchanges resolution with speed. This inherent characteristic depicted in Figure 3.6 is suited for multi-standard wireless receivers because by choosing different bandwidths and different sampling speeds, the same A/D converter exhibits different dynamic range and resolution. In the proposed multi-standard receiver, the A/D converter needs to perform as a 14bit-100KHz converter for DCS1800, a 11bit-700KHz converter for DECT and a 6bit-2.5MHz converter for WCDMA.

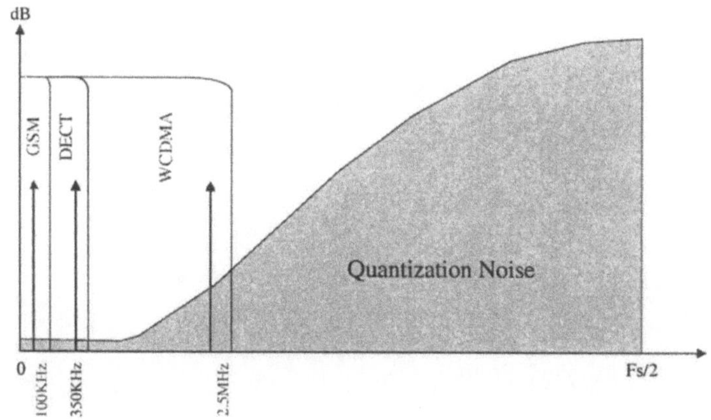

Figure 3.6. Inherent Characteristic of $\Delta\Sigma$ Modulator A/D Converter

5. Summary

A WCDMA/GSM/DECT multi-standard receiver is analyzed in a top-down manner in this chapter. First the multi-standard receiver architecture is presented. Design issues in the front-end, the base-band and the frequency synthesizer are discussed. The receiver specifications are then derived from the system requirements. Combined with the architecture, these specifications are distributed to the building blocks such that the shared hardware satisfies the requirement of all standards. Though we discuss the multi-standard receiver for WCDMA, GSM and DECT systems, the analysis method and the receiver architecture can be easily adapted to other multi-standard applications.

Chapter 4

RF CMOS BUILDING BLOCKS

RF CMOS circuit design techniques are essential for fully-integrated CMOS receivers. Design strategies for multi-standard LNAs are discussed. In this chapter, design techniques of two major building blocks in the RF parts are presented. For CMOS LNA design, an internal matching on-chip inductor is added between the input transistor and the cascode transistor. With this technique, both noise figure and voltage gain can be improved without degrading linearity performance. Frequency synthesizers for multi-standard operation are also discussed. A new prescaler architecture is proposed. Using this new architecture, a 2.5GHz prescaler using 0.35μm CMOS technology is realized with 4mW power consumption.

1. Multi-Standard Front-end

As the first on-chip stage to handle the received band, the LNA carries the heaviest burden in terms of noise and linearity. Since LNA is the first building block of the receiver chain, the noise figure of the receiver is dominated by LNA. Second, the LNA must match to 50Ω of the off-chip band filter's output impedance. Third, the LNA gain must be chosen according to the noise and linearity of the mixer. If this gain is too low, the mixer noise dominates the overall noise while if it is too high, the input signal to the mixer creates large intermodulation products. Thus noise, input matching and linearity requirements limit acceptable LNA topologies to only a few. Widely used are the common-gate and inductively-degenerated cascode topologies since they provide both a high reverse isolation (thus ensuring stability) and input resistance that can be set to 50Ω by design.

For a multi-standard receiver, there are several ways to implement the front-end LNAs. The straight-forward way is to use a wideband LNA, which performs the required specifications covering the overall bandwidth of all the standards.

An example of this approach is presented in [3], which targets both 900MHz and 1.9GHz bands operations. A dual-loop wideband active-feedback is applied to achieve the wideband input matching. The chip combines the LNA with the mixer without external interface. It uses a 25GHz bipolar technology and achieved 20dB conversion gain, +17dBm OIP3 and 3.8dB of NF. It consumes 13mA DC-current from 5V voltage-supply. Note that wideband operation allows more noise injection to the system.

A more power efficient low noise approach is to implement the multi-standard front-end by multiple LNAs, as it is shown in Figure 3.3. The work in [64] is such an example using BiCMOS technology. The paper presents a RF front-end for dual-band dual-mode operation. It is designed to be used in a zero-IF WCDMA and GSM receiver. The front-end has been fabricated in a 0.35-mum BiCMOS process with 1.8V supply voltage. It has two separate single-ended inputs, one input for each standard. This assumes there is no band-select filtering in the receiver [64]. With the exception of the input transistors and matching inductors of the LNA, all on-chip devices are utilized in both modes. The measured noise figure and voltage gain are 2.3dB, 39.5dB for the GSM and 4.3dB, 33dB for the WCDMA standards, respectively.

As depicted in Figure 3.3, we propose to implement the front-end by using three separate LNAs so that each LNA can be optimized in terms of power, noise, gain and linearity specifications. The following sections present a novel design approach to optimize a GHz CMOS LNA.

2. RF CMOS LNA Design

RF CMOS LNAs have been extensively investigated [47][69]. The cascode structure has been considered to be the best topology for an integrated low noise amplifier because it can satisfy requirements for both noise and power gain simultaneously [47]. However, the matching of the output load of the common-source stage is often less emphasized in the analysis of this topology. In order to obtain more accurate analysis, the loading effects should be considered carefully since the input-to-output isolation of the common-source stage is not good enough at GHz frequency IC. Our work shows the importance of the matching between the two transistors in the cascode structure. Dramatic performance improvement can be achieved by adding a matching inductor between the two stages.

2.1. A Novel Design Approach for LNAs

The simplified AC signal path of a cascode LNA is shown in Figure 4.1. Note that a matching inductor L_a is added between the common-source input stage M_1 and the common-gate stage M_2. The input source impedance and output load impedance of the LNA are assumed to be 50Ω.

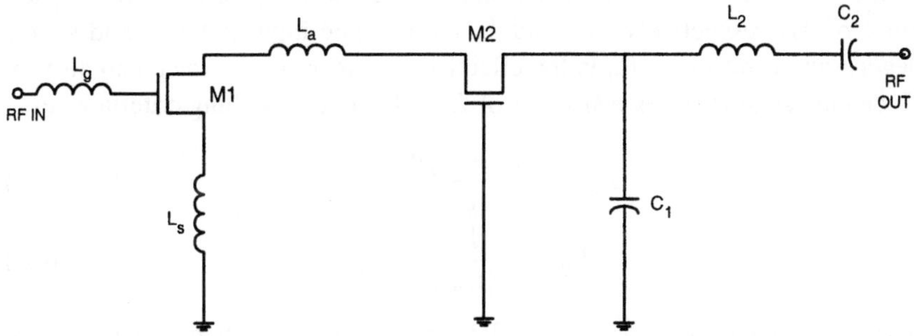

Figure 4.1. AC Signal Path of the Modified Cascode LNA

It is hard to find the matching network of the two-stage amplifier because of the mutual dependence of the input and output loading. However, the LNA can be divided into two parts in the design considering the high input-output isolation of the common-gate stage M_2, as shown in Figure 4.2. This means

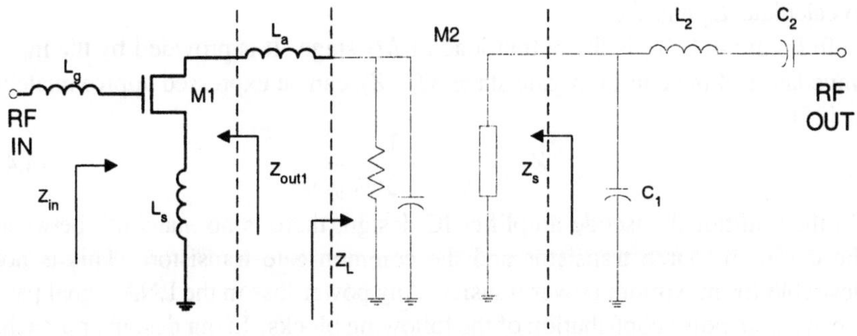

Figure 4.2. A Two-stage Design Matching Strategy

that the output matching networks of the common-gate stage can be separately designed without considering the input side of M_2. The input matching and output matching for the inductor degenerated common-source stage M_1 need to be considered simultaneously. Basically, the design approach can be divided into three steps. First, L_g and L_s are estimated by ignoring Miller effect of C_{gd1} of M_1. The design technique of this part is well-established [69][26][57]. The input impedance of the first stage is given by:

$$Z_{in} = j\omega(L_g + L_s) + \frac{1}{j\omega C_{gs1}} + R_g + \frac{g_{m1}}{C_{gs1}} L_s \qquad (4.1)$$

where g_{m1} and C_{gs1} are the transconductance and the gate-to-source capacitance of M_1 respectively. L_g and L_s are the gate input inductor and source degeneration inductor. R_g is the effective gate resistance, which can also include the parasitic series resistance of L_g. The input matching criteria gives:

$$L_s = \frac{(R_s - R_g)C_{gs1}}{g_{m1}} \tag{4.2}$$

$$L_g = \frac{1}{\omega_0^2 C_{gs1}} - L_s \tag{4.3}$$

where R_s is the source resistance (50Ω) and ω_0 is the operating frequency. The gate width of M_1, W_1, should be determined before the calculation of L_g and L_s. W_1 can be initially determined with the equations given in [69]. There are several design issues that should also be taken into account in determining the size of M_1. In Figure 4.1, L_a is implemented by an on-chip spiral inductor. L_s and L_g are realized by a combination of bonding-wire inductor and on-chip spiral inductor. Since the Q factor of the spiral inductor is very low, L_g is required to be as small as possible so that noise figure will not be degraded too much by its parasitic resistance. From the resonance equation 4.1, C_{gs1} should be large, which leads to a large W_1 size. After W_1 is chosen, it is straightforward to calculate L_g and L_s.

In Figure 4.2, Z_L is the output load of M_1 stage. It is provided by the input impedance of the common-gate stage M_2. Z_L can be expressed approximately as [25]:

$$Z_L = \frac{1}{g_{m2} + j\omega C_{gs2}} \tag{4.4}$$

In the traditional cascode amplifier IC design, there is no matching between the common-source transistor and the common-gate transistor. This is not desirable for maximum power transfer. Any power loss in the LNA signal path means more noise contribution of the following blocks. In our design approach, the cascode LNA is considered as a two-stage amplifier shown in Figure 4.2. The inter-stage matching is needed to provide the highest gain. Since the input impedance of the second stage (Z_L) and output impedance of $M_1(Z_{out})$ are both capacitive, a series spiral inductor L_a is required between the two transistors to improve matching. With inductor L_a, the gain of first stage is improved and thus the Miller effect of C_{gd1} becomes bigger. So the input matching criteria equation 4.2 does not hold anymore. Optimization should be performed by simulations to obtain new values of L_g, L_s and L_a. Because of the maximum power transfer, the input referred noise contribution of the common-gate stage M_2 can be greatly reduced and thus the overall noise figure of the LNA will be decreased. After the input stage is optimized and thus Z_L is determined, the design of the common-gate stage is to provide the accurate input impedance of Z_L. W_2 is chosen such that g_{m2} and C_{gs2} give the exact value of Z_L according to equation 4.4. A very small feedback capacitor C_1 between the drain and the gate of the common-gate stage makes the amplifier stable while keeping the gain almost the same. The final step is to design the output matching of the common-gate stage. L_b is an off-chip inductor connected to the power supply. It is ideally noiseless and allows a large signal swing at the output even with a low supply voltage. Along with the series capacitor C_2, the output impedance is matched to 50Ω.

2.2. A 2.0GHz LNA Design Example

A 2.0GHz CMOS LNA is designed using the presented method [42][79]. Figure 4.3 shows the schematic of the LNA. M_1 and M_2 are the cascode transistors. R_{b1}, R_{b2}, M_3 and M_4 provide biasing voltages for M_1 and M_2. C_3 is used to make the gate of M_2 AC grounded at the operating frequency. C_1 helps to stabilize the circuit. L_g and L_s are implemented by combination of on-chip spiral inductors and bonding-wire inductors. L_a is a fully on-chip inductor working as the inter-stage matching component. L_{out} and C_{out} form the output matching network. Because L_g, L_s and L_a are realized by on-chip spiral inductors, the parasitic resistance becomes detrimental to the noise performance. The value of L_s is around 2nH. It can be easily implemented by a

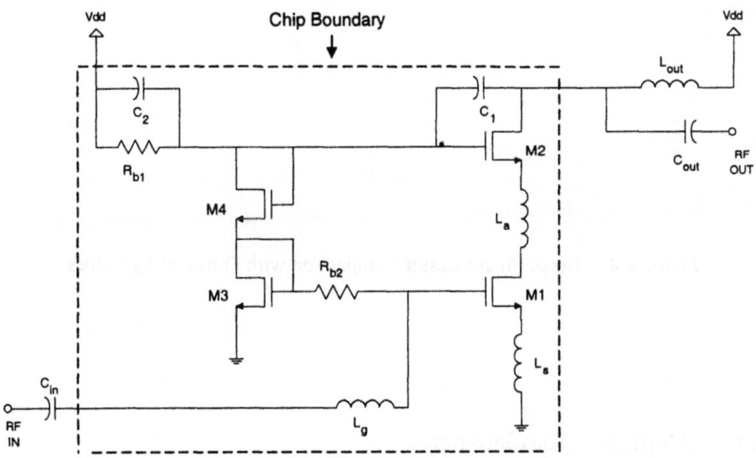

Figure 4.3. Schematic of the 2.0GHz Modified Cascode LNA

high Q bonding-wire inductor. According to equation 4.2, a large width W_1 is chosen to reduce the value of L_g, thus providing less noise degradation. Including L_a, the first stage amplifier M_1 is optimized with the load impedance of Z_L. According to equation 4.4, the width of M_2 is obtained as W_2. As described previously, L_a is implemented by an on-chip inductor. The parasitic resistor of the spiral introduces potential thermal noise that may degrade the total noise figure. On the other hand, the inter-stage gain increased by L_a improves noise performance even more. The overall noise performance is better. Figures 4.4 and 4.5 show simulation results by sweeping L_a. Note that the model of all the spiral inductors in the schematic are extracted from layout, thus the parasitic resistances are taken into account. By adding a 5nH inductor, the gain is improved by 3dB and noise figure is decreased from 3.0dB to 2.0dB. The final simulation result is shown in Table 4.1.

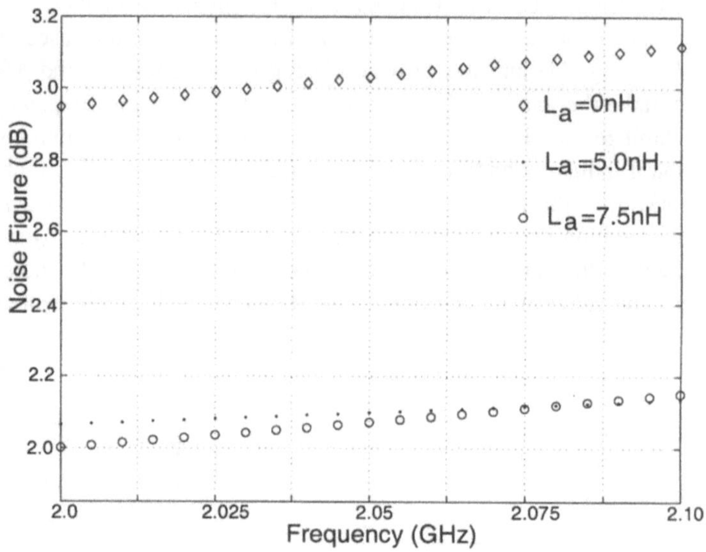

Figure 4.4. Noise Figure Result Comparison with Different L_a Values

Table 4.1. 2.0GHz LNA Simulation Result

Process	0.35μm CMOS
Supply Voltage	3V
Frequency Range	2.0GHz-2.1GHz
Noise Figure	2.0dB
Power Gain	19.7dB
IIP_3	5dBm
S_{11}	-15dB
Current	3mA

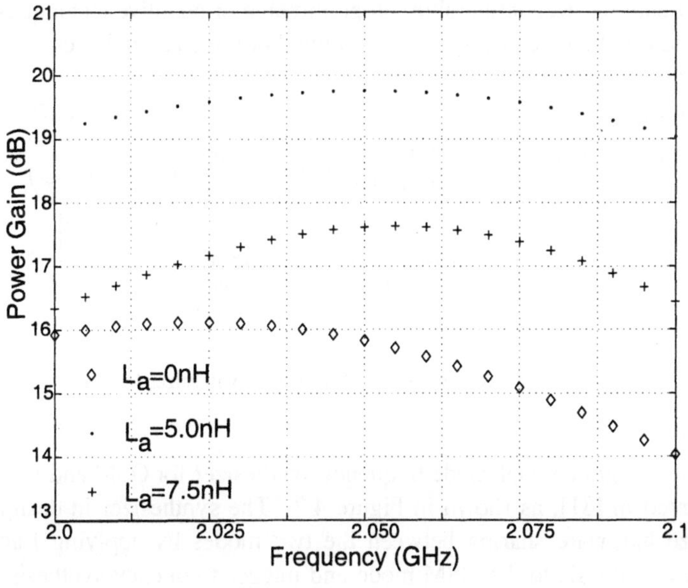

Figure 4.5. Power Gain Result Comparison with Different L_a Values

The result shows that by adding an interstage matching inductor L_a, the gain and noise performance are improved while third-order nonlinearity is kept high.

3. Multi-Standard Frequency Synthesizers

The PLL-type frequency synthesizer is one of the major building blocks for RF transceivers. Figure 4.6 shows the basic architecture of a PLL frequency synthesizer. For a frequency synthesizer in a wireless receiver, the major specifications include output frequency range, frequency resolution, phase noise, settling time, frequency accuracy and etc.. There are several well-known tradeoffs in the synthesizer design. For example, in a high resolution frequency synthesize it is difficult to achieve fast frequency settling behavior. Low phase noise output spectrum can be obtained by designing a wideband loop filter. However, a wideband loop filter would pass more spurious signals from the phase detector and generate more spurious spectrum by FM modulation. Optimization is always needed in frequency synthesizer design, especially the loop filter design. For each standard, there is a set of these specifications, thus parameters, such as loop filter bandwidth, VCO sensitivity, charge pump gain and divider ratio must be considered in the context of a multi-standard design. For a multi-standard receiver, these specifications get much more complex. It is very difficult, if not impossible, to design one optimized synthesizer satisfying all specifications. Reconfigurable, programmable or adaptive techniques must be included in the block design, as we pointed out in chapter 3 section 2.3.

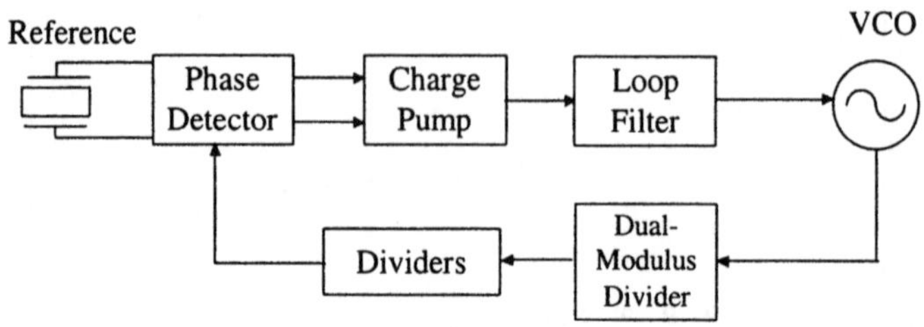

Figure 4.6. Phase-locked Loop Frequency Synthesizer Diagram

A fully-integrated dual-mode frequency synthesizer for GSM and WCDMA is presented in [81], as shown in Figure 4.7. The synthesizer in designed to maximize hardware sharing between the two modes by applying fractional frequency synthesis to the GSM mode and integer frequency synthesis to the WCDMA mode. A dual-mode VCO is also proposed for the enhanced tuning range with an accumulation mode NMOS varactor for band-to-band tuning and a p^+N junction varactor for in-band tuning.

In the following part, we will discuss one challenging building block in PLL frequency synthesizer design using CMOS technology. That is the high-speed prescaler design.

4. GHz Prescaler

Besides the voltage-controlled oscillator, dual-modulus divider (prescaler) is the other building block that operates at the highest frequency range in the synthesizer. It divides the VCO output frequency by a certain ratio to convert it to a low-frequency signal. The prescaler operates at two different divide ratios, and is controlled by a modulus control signal. Compared with fixed-division-ratio frequency dividers, extra dual-modulus control logic is needed in a prescaler. As a result, the operating speed is slowed down. Thus high-speed dual-modulus prescalers are more difficult to design than fixed-division-ratio frequency dividers and normally they are much slower than fixed ones.

4.1. Prescaler Architectures

A traditional high-speed 32/33 dual-modulus prescaler [33] generally consists of a synchronous divide-by-4/5 counter and an asynchronous divide-by-8 counter as shown in Figure 4.8. The divide-by 32/33 function is selected through a modulus-selection input M. With this input at logic zero the syn-

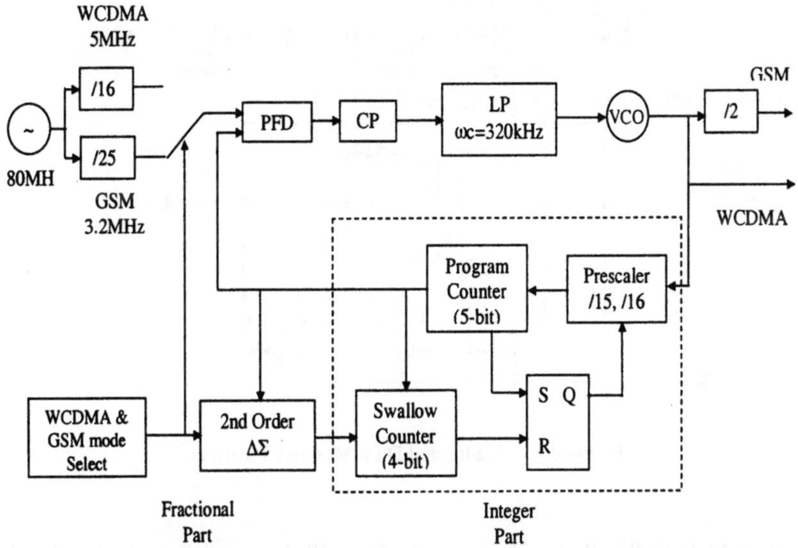

Figure 4.7. Block Diagram of the Dual-mode Frequency Synthesizer

chronous counter does divide-by-4 only, and the prescaler functions as a divide-by-32 counter. For the divide-by-33, the synchronous counter does divide-by-4 until all the flip-flops in the asynchronous counter reset to logic zero. Then the outputs from these flip-flops, together with the modulus-selection input at logic one, generate a signal pulse enabling the synchronous counter to do divide-by-5 only once in a 33-clock cycle. Afterwards the divide-by-4 is resumed. In this architecture, the prescaler speed is limited by the speed of the synchronous counter, implemented using three D-type flip-flops and two NAND gates.

One way to increase the operation frequency is to design faster D-type flip-flops. This has been investigated extensively [89][27][9][88]. The state-of-the-art fast D-type flip-flop topology is presented in [27]. Small modification [35] at the architecture level to lower the power optimization is to replace the divide-by-4/5 synchronous counter with a divide-by-2/3 counter so that only one D-FF is clocked by the high-frequency input signal. This also reduces the capacitance load on the clock buffer. However, the disadvantage of this configuration is a lower operation margin compared with the conventional configuration and thus the operating frequency is decreased [35].

Another prescaler architecture is proposed and implemented in [11]. It consists of a chain of seven pure divide-by-two circuits interrupted by a phase-select block. Since the first stage is only a toggle-flip-flop, input frequencies as high as asynchronous dividers can be obtained. The dual-modulus operation is based on the 90-degree phase relationship between the outputs of the master and the

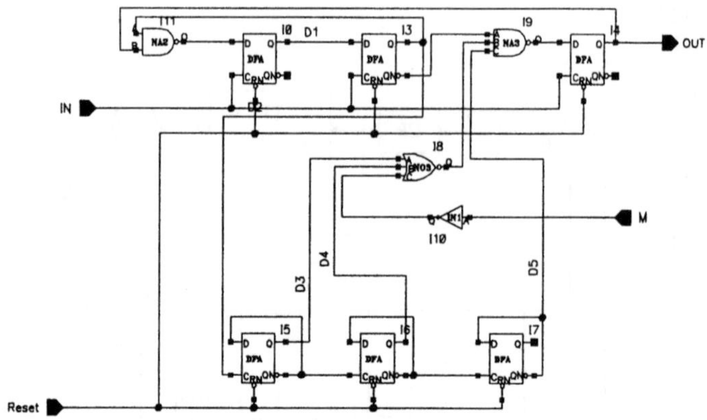

Figure 4.8. Traditional Dual-Modulus Prescaler

slave of M/S D-flip-flop. To generate four 90-degree different signals, the differential signals of the VCO are needed.

The new dual-modulus prescaler topology proposed here is used for single-ended input and the phase-select block is greatly simplified. The block diagram is shown in Figure 4.9. It consists of one full speed D-FF, a half-speed buffer,

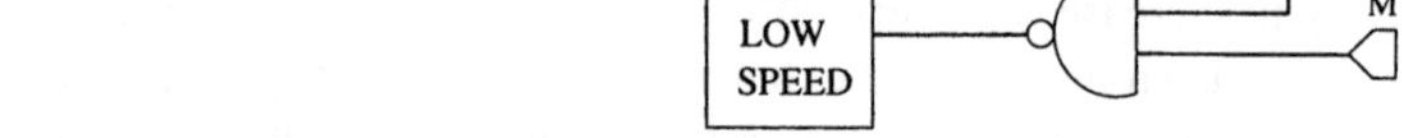

Figure 4.9. A New Architecture of 32/33 Prescaler

phase switching block and a chain of five pure divide-by-2 circuits. The new prescaler operates as follows. The single ended input signal IN is fed to the full-speed divide-by-two flip-flop. A buffer and an inverter are used to generate two half-speed signals with the phase difference of 180 degree. When the mode control input M is low, the low speed divide-by-2 DFF is disabled, thus the phase-switch control signal will be constant and the switch simply stays at the original state connecting one of the two inputs to the divide-by-16 counter. The resulting divide-ratio of the prescaler is $2 \times 16 = 32$. The divide-by-33 operation is enabled by setting M high. The divide-by-2 low speed DFF is now working. On every positive edge of the OUT signal, the output of the DFF will flip from zero to one, or from one to zero. Thus the input of the divide-by-16 asynchronous divider is switched to connect one of the two different signals with 180 phase shift accordingly. The signal is thus delayed by 180 degree once OUT has one positive edge. So the output period is increased with this delay which equals one period of the IN signal. The overall prescaler division ratio now is $32 + 1 = 33$.

Based on the new topology, a CMOS 32/33 prescaler used in a 2.4GHz frequency synthesizer for ISM band wireless transceivers was designed in a $0.35\mu m$ digital CMOS technology and 2.7V supply voltage.

4.2. High-speed D-type Flip-Flop (DFF)

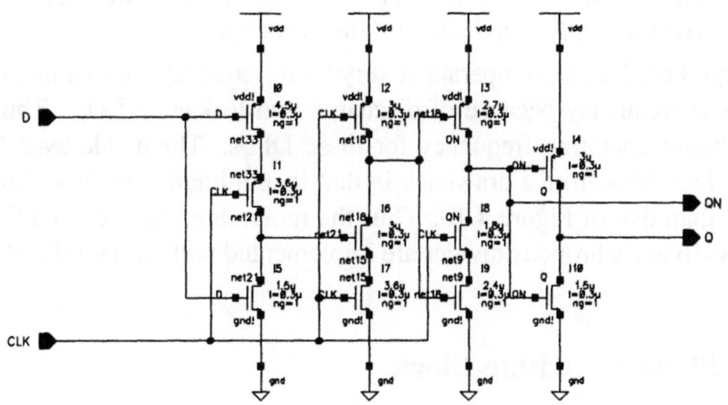

Figure 4.10. Yuan and Svenssan DFF

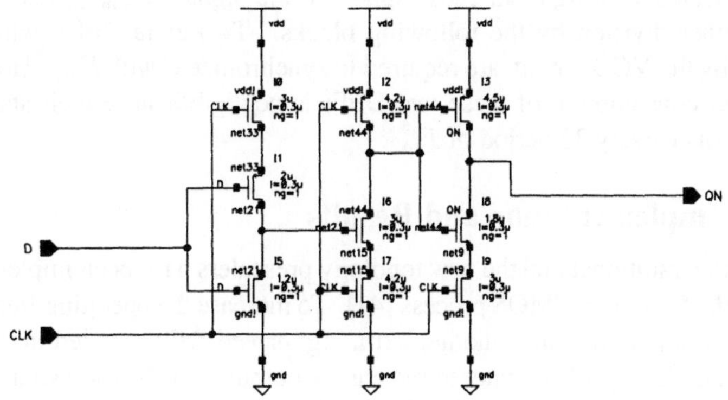

Figure 4.11. Huang's DFF

Yuan and Svenssan developed a nine-transistor DFF without feedback, shown in Figure 4.10. This data-inverting DFF is driven by a true single-phase clock avoiding clock inverters and explicit slave stages usually found in standard configurations. It is constructed by three stages and the input data is latched by the positive transition of the clock signal [89]. Huang proposed an even faster D-FF topology in [27], shown in Figure 4.11. Though the difference in circuit configuration seems to be a minor one, it is found that 20% higher speed is consistently achievable for the flip-flop in [89]. The CMOS implementation of M/S ECL D-flip-flop is an alternative way to realize high-speed divide-by-2 counter [11]. The circuit achieves the speed enhancement by a reduction in voltage swing from the input to output. Simulation indicates a speed improvement over a standard implementation. However, the sizing of the DFF in [11] is not easy and care has to be taken to ensure proper operation with the VCO and the following logic circuits. Considering low voltage and low power applications, Huang's dynamic D-FF is undoubtedly the best choice.

Though these DFFs can operate at very high frequency, they cannot operate at very low frequency because of discharge and leakage effects. Thus there is a minimum operating frequency for these DFFs. The divide-by-2 DFF in Figure 4.11 suffers from a drawback in that its minimum operating frequency is higher than that of Figure 4.10. Thus the remaining low speed DFFs in the divide-by-16 asynchronous divider are implemented with the circuits of Figure 4.10.

4.3. Phase Switching Block

The phase switching block is shown in Figure 4.12. The alternating selection of the two different signals is achieved by control signal C. Note that the input signal is buffered and inverted by two inverters. A very important aspect of this circuit is what happens at the transfer between the two signals with different phases. Careless design can cause spikes in the signal of F_{out2}, resulting in an improper division by the following blocks. Two extra DFFs, which are clocked by the VCO output, are required to synchronize C with F_{in2}. However, the power consumption of these two DFFs is negligible since their states are changed once every 32 period of F_{in2}.

4.4. Implementation and Results

Both the traditional and the new topology prescalers have been implemented in a standard $0.35\mu m$ CMOS process [41]. To increase the operating frequency of the traditional prescaler, another extra high-speed DFF is added in the synchronous divide-by-4/5 counter to reduce the output load. It is shown in Figure 4.13. The prescaler benefits from this modification by around 20% speed increase according to simulation, from 1.7GHz to 2.1GHz. Figure 4.14 shows

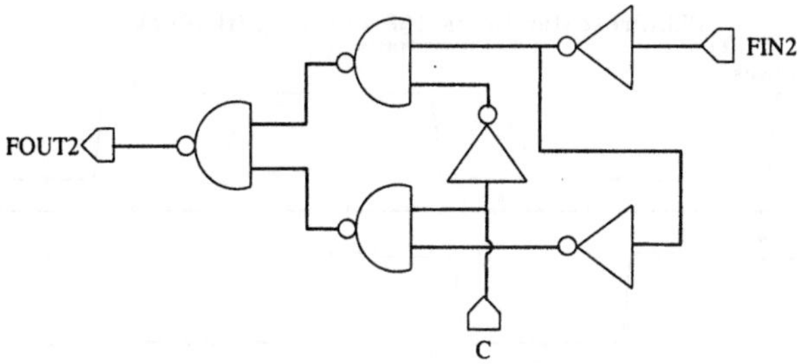

Figure 4.12. Phase Switch Block

the output waveforms of the prescaler designed. Table 4.2 summarizes the performance characteristics of the two circuits. It clearly demonstrates the superiority of the proposed architecture. The proposed new prescaler is fabricated in $0.35\mu m$ CMOS technology. The measured waveform is shown in Figure 4.15 and 4.16. The supply voltage is 2.7V with current consumption of 1.5mA.

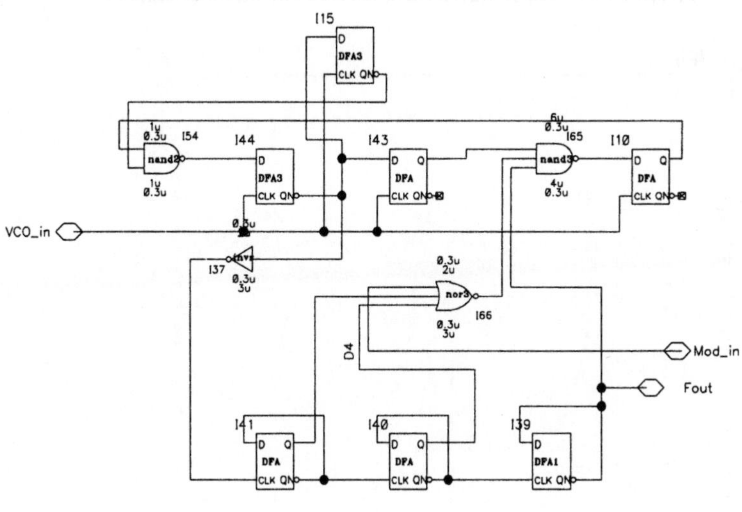

Figure 4.13. Modified Traditional Prescaler Architecture

5. Summary

For a fully-integrated CMOS multi-standard receiver, RF circuits are the most challenging parts. The relatively low performance of CMOS RF circuits

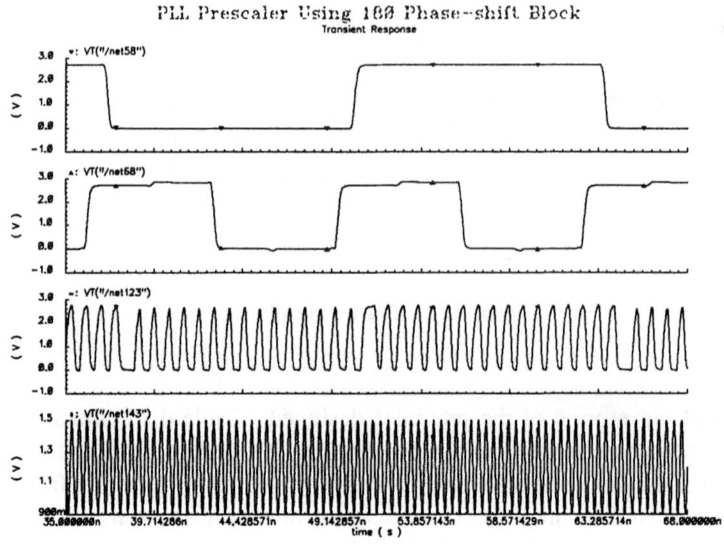

Figure 4.14. Output Waveform of the New Architecture Prescaler

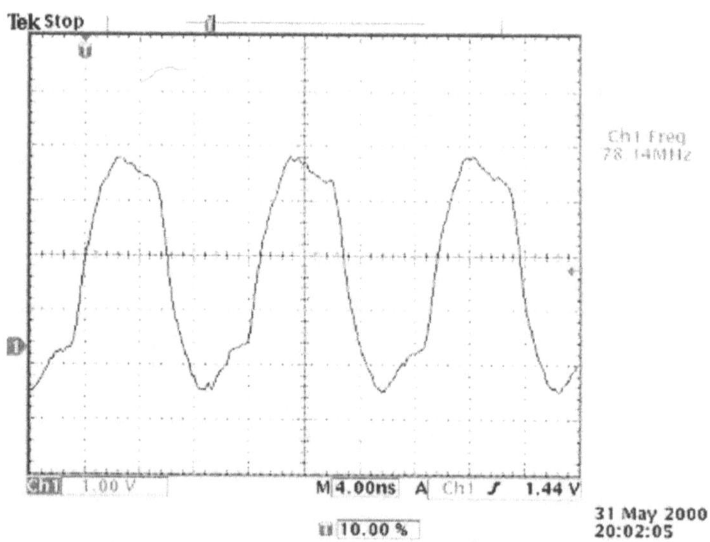

Figure 4.15. Measurement Output Waveform of the New Architecture Prescaler: F_{in}=2.500GHz, N=32 and F_{out}=78.1MHz

compared with their Bipolar counterpart is due to both the low g_m of transistors and the lack of accurate modeling for MOS transistors and passive components,

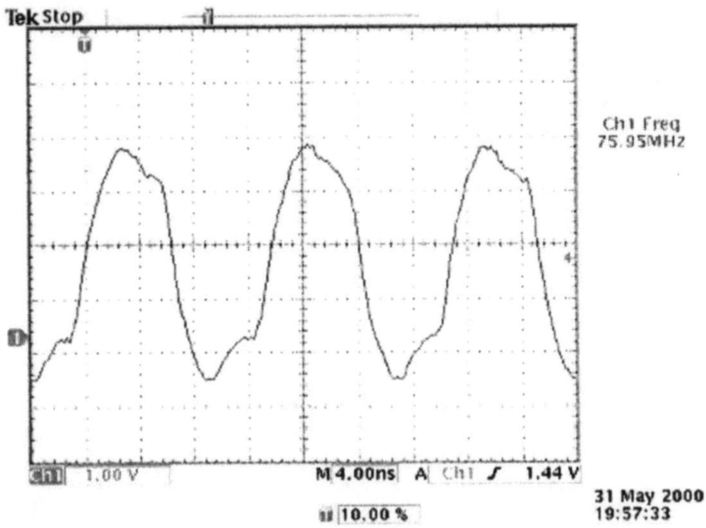

Figure 4.16. Measurement Output Waveform of the New Architecture Prescaler: F_{in}=2.500GHz, N=33 and F_{out}=75.9MHz

Table 4.2. Prescalers Simulation Performance Comparison

Specification	Traditional Topology	New Architecture
Highest Operating Frequency	2.1GHz	3.0GHz
Supply Voltage	2.7V	2.7V
Current Consumption	1.5mA @ 2.1GHz	1.5mA @ 3.0GHz
Input Amplitude	1.6Vpp	0.6Vpp

e.g. on-chip inductors. Thus smart design techniques at both the circuit and system levels are necessary. In this chapter, we briefly discussed the design issues of front-end LNAs and the frequency synthesizer design for multi-standard applications. A novel design approach for LNAs and a high-speed prescaler using CMOS single-ended DFF operating up to 2.5GHz are presented.

Chapter 5

MULTI-STANDARD $\Delta\Sigma$ MODULATOR: ANALYSIS

This chapter reviews some of the fundamental issues in the design of $\Delta\Sigma$ modulators. The basic concept of a $\Delta\Sigma$ modulator is described and the linearized models are reviewed. Following this basic introduction, tradeoffs among a variety of $\Delta\Sigma$ architectures suitable for multi-standard applications are explored. A programmable $\Delta\Sigma$ A/D converter is proposed. The detail system level analysis and simulation are presented to obtain the specifications for each building block in the modulator.

The programmable modulator targets WCDMA, GSM and DECT operations. It behaves as 14bit, 12bit and 7bit A/D converter for GSM, WCDMA and DECT signal respectively. This chapter covers the analysis from deciding the order and OSR (oversampling ratio) of the modulator to determining all the specifications of the building blocks, such as OTAs (operational transconductance amplifier), comparators and internal DAC. Based on these results, the design of a switched-capacitor modulator is straightforward, and is presented in the next chapter.

1. Quantization Noise and Noise-shaped $\Delta\Sigma$ Modulator

Once sampled, the signal samples must also be quantized in amplitude to a finite set of output value. The difference between the original continuous amplitude and the new "mapped" value represents the quantization error. The number of levels is proportional to the resolution of the quantizer used in the ADC. It can be observed that increasing the quantizer resolution will decrease the quantization error.

The error is completely defined by the input, but if the input changes randomly between samples by amounts comparable with or greater than the threshold spacing, without causing saturation, then the error is largely uncorrelated from sample to sample and has equal probability of lying anywhere in the range $\pm\Delta/2$. Further, if it is assumed that the error has statistical properties, which are independent of the signal, the error can then be represented by noise. The quantization noise is given by the mean-square value of the quantization error described above.

$$P_Q = \frac{1}{\Delta}\int_{\frac{-\Delta}{2}}^{\frac{\Delta}{2}} q^2 dq = \frac{\Delta^2}{12} \tag{5.1}$$

When the quantized signal is sampled at f_S, all of its power folds into the frequency band from 0 to $f_s/2$. Thus the higher the sampling frequency is, the lower the quantization noise spectrum density becomes. Each doubling of the sampling frequency decreases the in-band quantization noise by 3dB, increasing the ADC resolution by half bit.

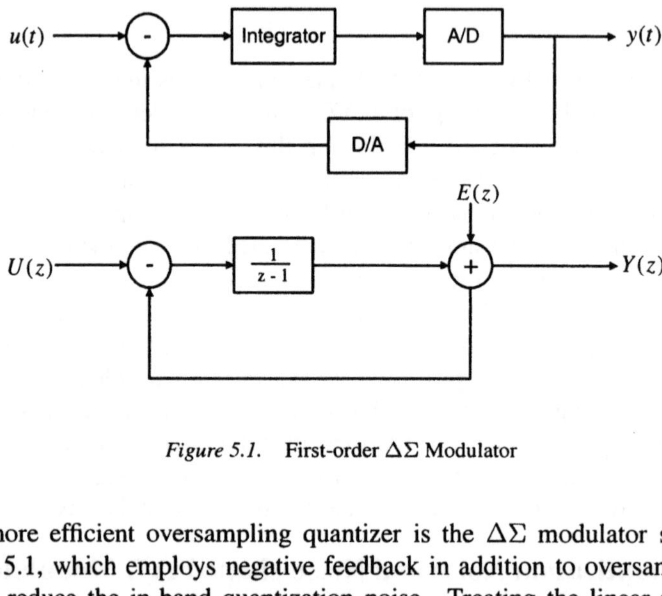

Figure 5.1. First-order ΔΣ Modulator

A more efficient oversampling quantizer is the ΔΣ modulator shown in Figure 5.1, which employs negative feedback in addition to oversampling to further reduce the in-band quantization noise. Treating the linear model as having two independent inputs, we can derive a signal transfer function, $S_{TF}(z)$, and a noise transfer function, $N_{TF}(z)$.

$$S_{TF}(z) = \frac{Y(z)}{U(z)} = \frac{H(z)}{1+H(z)} \tag{5.2}$$

$$N_{TF}(z) = \frac{Y(z)}{E(z)} = \frac{1}{1+H(z)} \tag{5.3}$$

To noise-shape the quantization noise in a useful manner, $H(z)$ is chosen such that its magnitude is large from 0 to f_0 (i.e. over the frequency band of interest). With such a choice, the signal transfer function will approximate unity over the frequency band of interest very similarly to an opamp in a unity-gain feedback configuration. Furthermore, the noise transfer function will approximate zero over the same band. Thus, the quantization noise is reduced over the frequency band of interest while the signal itself is largely unaffected. The high-frequency noise is not reduced by the feedback as there is little loop gain at high frequencies. However, additional post filtering, which is realized by a digital decimation filter, can remove the out-of-band quantization noise with little effect on the desired signal.

We can obtain first-order noise shaping by letting $H(z)$ be a discrete-time integrator [34]. Specifically,

$$H(z) = \frac{1}{z-1} \tag{5.4}$$

The noise transfer function is derived assuming $f_0 << f_S$ as

$$|N_{TF}(f)| = 2\sin(\frac{\pi f}{f_S}) \approx \frac{2\pi f}{f_S} \tag{5.5}$$

Now the quantization noise power over the frequency band from 0 to f_0 is given by:

$$P_e = \int_{-f_0}^{f_0} (\frac{\Delta^2}{12})^2 \frac{1}{f_S} (\frac{2\pi f}{f_S})^2 df = \frac{\Delta^2 \pi^2}{36} (\frac{1}{OSR})^3 \tag{5.6}$$

where OSR is the over-sampling ratio. Then the maximum SNR from this case is given by:

$$SNR_{max} = 10log(\frac{P_s}{P_e}) = 6.02N - 3.14 + 30log(OSR) \tag{5.7}$$

We see here that doubling the OSR gives an SNR improvement for a first-order modulator of 9dB, equivalently, a gain of 1.5bits/octave. This result should be compared with the 0.5bits/octave when oversampling with no noise shaping. This noise shaping effect can be generalized for the in-band quantization in an ideal L-order ΔΣ modulator. The dynamic range (with quantization noise being the only noise source) can be easily derived as:

$$DR = \frac{3}{2} \frac{2L+1}{\pi^{2L}} G^2 M^{2L+1} (2^B - 1)^2 \tag{5.8}$$

where B is the quantizer resolution in bit; L is the order of the ΔΣ modulator; M is the oversampling ratio and G is the overall gain in the forward path from the input to the quantizer. For every doubling of the oversampling ratio, the dynamic range increases by 3(2L+1)dB or (L+0.5) bits. However the higher-order modulator (3rd-order or higher) may experience limit-cycle oscillations or instability very easily. Certain techniques are needed to overcome this problem.

2. Multibit and Cascaded ΔΣ Modulators

Second-order ΔΣ modulators are widely used because they are simple to implement and insensitive to component mismatch. The problem with a second-order modulator is the high oversampling ratio, thus high sampling frequency, required to meet a very high dynamic range requirement. Thus a single second-order modulator is impractical for high-speed high-resolution applications. To increase the order of the modulator, the designer must pay significant attention to stability. The designer need scale the integrator gains and place clippers on the outputs of each of the integrators to assure that the modulator remains stable when overloaded and during power up.

One solution to the above problem is to use a multibit quantizer in the modulator loop. The primary advantage of noise-shaping modulator employing multibit quantizers is that the ratio of the total quantization noise power to the signal power at the modulator output is dramatically reduced from that of a 1-bit modulator, typically 6dB per each additional bit. Therefore, we can increase the overall resolution of any oversampling ratio, simply by increasing the number of levels in the internal data converters. Equivalently, a multibit noise-shaping modulator can achieve a resolution comparable to that of a single-bit modulator at a lower sample rate. This performance increase can be a significant advantage in high bandwidth applications. Another advantage of the lower clock rate possible with multibit modulators is the decreased power consumption in the digital circuitry.

A notable disadvantage of multibit systems is that they lack the ability of single-bit systems to achieve excellent linearity without the use of matched components. The integral linearity of a noise-shaping conversion system is not better than the integral linearity of the multibit internal DAC. Therefore, achieving high integral linearity and low total harmonic distortion (THD) appears to require precisely matched components. A secondary disadvantage of multibit modulators is that more analog circuitry, which is generally more difficult to design than digital circuitry, is required.

Another approach for realizing high-resolution modulators is to use a cascade-type structure where the overall higher-order modulator is constructed using lower-order ones. The lower-order ones are normally first-order or second-order modulators, which are inherently stable. Since stability criteria are relaxed as compared to the higher-order loops, the cascade modulators approach the dynamic range bound in equation 5.8 more closely than higher-order single loop implementations.

Take a two-stage modulator as an example. The cascaded modulator is not a single-loop system. The quantization error from the first modulator is converted by the second modulator. The output of the second modulator is digitally recombined with the first modulator output to cancel the first modulator error. The degree of cancellation depends on how well the analog implementation of the transfer function matches the digital implementation of the inverse of the transfer function. The performance of the cascaded modulator is more sensitive to the imperfection of the analog components than that of the single-loop modulator. However, the accuracy requirement on the analog components is nowhere near as severe as for an equivalent successive approximation converter, because the in-band noise has been greatly reduced by noise shaping prior to the digital cancellation. Note that arbitrarily low quantization noise cannot be obtained in practice by constructing very high order modulators. At some point the performance will be limited by uncancelled noise from the first modulator. When that point is reached, no gain in performance will be realized by adding more stages to increase the overall modulator order.

The cascade architecture can be combined with multibit D/A converters to improve the dynamic range further [74]. To avoid sensitivity to the precision of the DAC, the multibit quantizer can be placed in the final stage of the cascaded modulator. The more critical first-stage quantizer has only two analog output levels and is therefore inherently linear. The input to the second stage is the difference between the output and the input of the first-stage quantizer. That is, the input to the second stage is the quantization error of the first stage.

3. Programmable 2-2 MASH ΔΣ Modulator Architecture

The expression for the dynamic range, or SQNR (signal to quantization noise ratio) of a ΔΣ modulator is rewritten here.

$$DR = \frac{3}{2}\frac{2L+1}{\pi^{2L}}G^2 M^{2L+1}(2^B - 1)^2 \tag{5.9}$$

where B is the quantizer resolution in bits; L is the order of ΔΣ modulator; M is the oversampling ratio and G is the overall gain in the forward path from the input to the quantizer. Based on Equation 5.9, Table 5.1 shows the sampling frequency, order of the modulator and the corresponding achieved performance for different applications. From this table, we can get some interesting insight on the ΔΣ modulator programmability for multi-standard A/D converters.

- Low order ΔΣ ($L = 1$ and $L = 2$) cannot be used because of the very high sampling rate of 800MHz needed for high-resolutions application (GSM mode).
- High order ($L = 5$ and $L = 6$) does not show their advantages because for lower resolution, wide bandwidth standards, such as DECT and WCDMA, the modulator still needs higher sampling rate at low oversampling ratio.
- $L = 3$ and $L = 4$ are excellent for multi-standard wireless applications, because they can offer satisfied resolution and bandwidth for different standards with reasonable sampling rates.

Table 5.1. $\Delta\Sigma$ Modulator Architecture Consideration: Order and OSR

Sampling Rate for Bandwidth of 100kHz (GSM Family)							
Resolution(DR)		L=1	L=2	L=3	L=4	L=5	L=6
8bits(50dB)	Sampling Frequency	12.8MHz	6.4MHz	3.2MHz	1.6MHz	1.6MHz	1.6MHz
	OSR	60 64	17 32	10 16	8 8	6 8	5 8
10bits(62dB)	Sampling Frequency	51.2MHz	6.4MHz	3.2MHz	3.2MHz	3.2MHz	1.6MHz
	OSR	151 256	29 32	15 16	10 16	8 16	6 8
12bits(74dB)	Sampling Frequency	102.4MHz	12.8MHz	6.4MHz	3.2MHz	3.2MHz	3.2MHz
	OSR	380 512	50 64	22 32	13 16	10 16	8 16
14bits(86dB)	Sampling Frequency	204.8MHz	25.6MHz	6.4MHz	6.4MHz	3.2MHz	3.2MHz
	OSR	956 1024	87 128	32 32	18 32	13 16	10 16
16bits(98dB)	Sampling Frequency	819.2MHz	51.2MHz	12.8MHz	6.4MHz	6.4MHz	3.2MHz
	OSR	2400 4096	152 256	48 64	25 32	17 32	12 16

Sampling Rate for Bandwidth of 700kHz (DECT)							
Resolution(DR)		L=1	L=2	L=3	L=4	L=5	L=6
8bits(50dB)	Sampling Frequency	89.6MHz	44.8MHz	22.4MHz	11.2MHz	22.4MHz	11.2MHz
	OSR	60 64	17 32	10 16	8 8	6 8	5 8
10bits(62dB)	Sampling Frequency	358.4MHz	44.8MHz	22.4MHz	22.4MHz	22.4MHz	11.2MHz
	OSR	151 256	29 32	15 16	10 16	8 16	6 8
12bits(74dB)	Sampling Frequency	716.8MHz	89.6MHz	44.8MHz	22.4MHz	22.4MHz	22.4MHz
	OSR	380 512	50 64	22 32	13 16	10 16	8 16
14bits(86dB)	Sampling Frequency	1433.6MHz	179.2MHz	44.8MHz	44.8MHz	22.4MHz	22.4MHz
	OSR	956 1024	87 128	32 32	18 32	13 16	10 16

Considering the nonideality of practical implementations of $\Delta\Sigma$ modulators, e.g., the thermal noise and flicker noise in the circuits, choosing $L = 4$ is more reasonable than $L = 3$.

It is also shown that a 3-bit quantizer will offer 16dB (2.5bit) dynamic range improvement. This gives us more flexibility and improvement on the programmability.

Figure 5.2 shows the proposed programmable $\Delta\Sigma$ modulator architecture. The architecture is a 2-2 cascade switched-capacitor modulators. Each stage is a second-order modulator in order to guarantee stable operation with a full-swing input. To improve the dynamic range of the WCDMA/DECT operating modes, a 3-bit (more accurately is 9-level) ADC and DAC feedback is used as the second stage quantizer. Ideally the quantization noise can be decreased by 18dB. In DCS1800 mode, the 1-bit quantizer is used at the second stage to save power. As the resolution in DCS1800 mode is very high, the multibit DAC does not improve the performance much because the mismatch between analog and digital coefficients becomes dominant on the quantization noise floor. Thus switching off the 3-bit feedback DAC in DCS1800 mode is reasonable in terms of saving power without degrading performance. The operating modes are controlled by static switches. All the building blocks can be shared for three

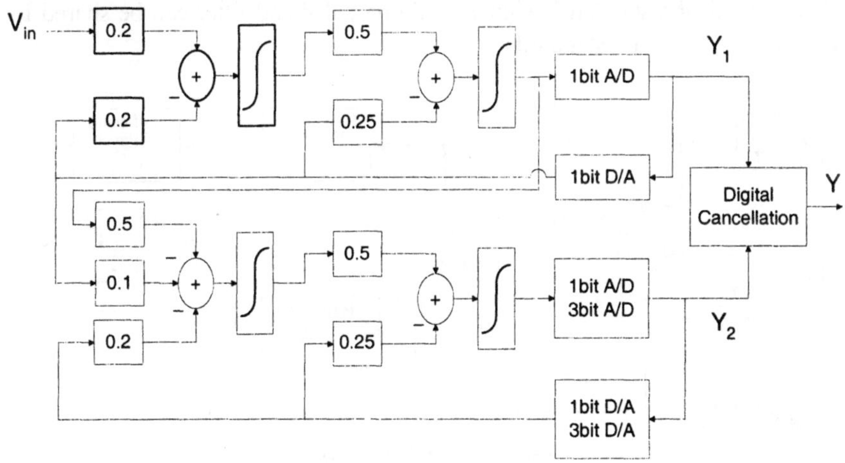

Figure 5.2. Programmable ΔΣ Modulator Diagram

modes, saving both chip area and power consumption. The outputs of the 1-bit and 3-bit quantizers are encoded in 2's complement before applying them to digital error cancellation logic. The frequency of the sampling clock is 13MHz, 26MHz and 39MHz for DCS1800, DECT and WCDMA respectively. Together with the second-stage quantizer switching, this guarantees that enough dynamic ranges are obtained by the ADC for the signals of DCS1800, DECT and WCDMA, which are 90dB, 75dB and 46dB, respectively. Other techniques are used to further decrease the power consumption. Capacitor scaling in the first stage integrator can greatly reduce the power consumption for DECT and WCDMA while the same OTA is used. The fourth stage integrator's gain coefficient is switched so that the range of the 3-bit quantization can be fully exploited, thus getting high resolution for DECT and WCDMA and saving power.

After the analog modulator, a programmable decimation filter is needed to complete the Nyquist A/D conversion. The architecture for the decimation filter is shown in Figure 5.3. It is a multi-stage decimation filter. It consumes much less power and saves considerable chip area compared with the one-stage realization. For DCS1800 standard, the oversampling ratio of the ΔΣ modulator is 64. The 13Msps signal is downsampled by two Sinc filter and two half-band filters. For DECT standard, the oversampling ratio is 16. The 26Msps signal is downsampled by one Sinc filter and two half-band filters. For WCDMA, the oversampling ratio is 8. The 39Msps signal is downsampled by one Sinc filter and one half-band filter. In Figure 5.3, one Sinc filter can be shared by all the standard's signals and one half-band filter can be shared by DECT and WCDMA standards.

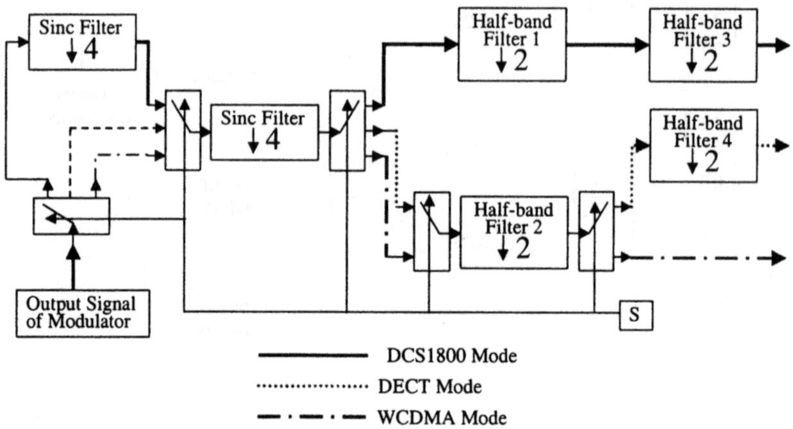

Figure 5.3. Programmable Decimation Filter Diagram

Shown in Figure 5.4 is the linear approximation of the 2-2 cascade modulator, where the quantizers are modeled by signal-independent additive error sources, while the integrators are represented by their transfer functions in the z-domain. Here, $E_1(z)$ and $E_2(z)$ model the quantization error of the first- and second-stage ADCs, respectively. The term $E_2(z)$ also contains a representation of nonlinearity in the second-stage ADC, while $E_D(z)$ models the error from nonlinearity in the multibit D/A converter in the second stage. An error source corresponding to $E_D(z)$ does not appear in the first stage because of the inherent linearity of the 1-bit D/A converter.

For the linearized model of Figure 5.4, the z-transform of the output of the first stage is:

$$Y_1(z) = z^{-1}X(z) + (1 - z^{-1})^2 E_1(z) \tag{5.10}$$

Thus the output of the first stage includes the input to the modulator delayed by one sample period plus the second-order difference of the first-stage quantization error $E_1(z)$. The input to the second stage is $E_1(z)$ and the transform of the second-stage output is

$$Y_2(z) = z^{-1}[E_1(z) - E_D(z)] + (1 - z^{-1})^2 E_2(z) \tag{5.11}$$

The error cancellation logic combines the digital outputs from the two stages according to:

$$Y(z) = z^{-1}Y_1(z) - (1 - z^{-1})^2 Y_2(z) \tag{5.12}$$

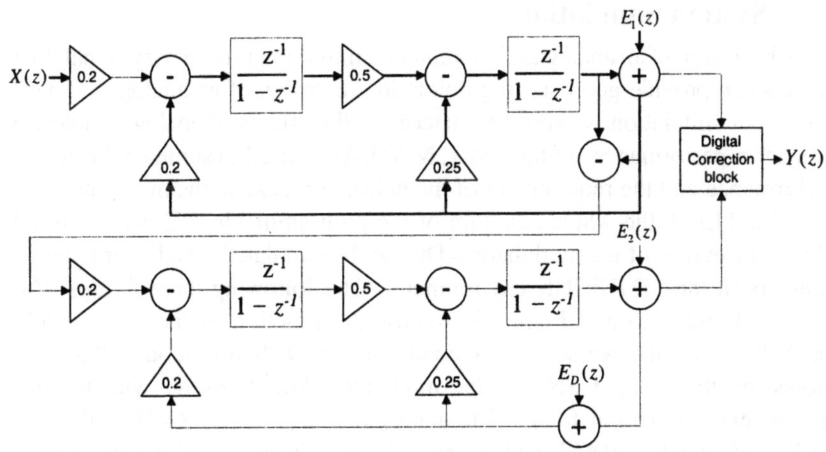

Figure 5.4. **Linear Model of the Cascaded Multibit ΔΣ Modulator**

so as to cancel the quantization error $E_1(z)$ of the first stage. The resulting output of the overall modulator is obtained by substituting (5.10) and (5.11) into (5.12),

$$Y(z) = z^{-2}X(z) + z^{-1}(1 - z^{-1})^2 E_D(z) - (1 - z^{-1})^4 E_2(z) \qquad (5.13)$$

Thus, ideally the quantization error of the first stage is cancelled and the quantization error of the second stage is attenuated in the baseband by fourth-order shaping. As in the cascade of two second-order stages, fourth-order noise shaping is achieved without instability because the constituent stages are independently stable. Also because the quantization error of the second stage originates from a multibit quantizer, the modulator's dynamic range is improved by an increase in the quantizer resolution accordingly. In contrast to a single-stage multibit modulator, the error resulting from DAC nonlinearity, $E_D(z)$, does not enter this modulator at its input but instead is attenuated in the baseband by second-order shaping. The second-order shaping makes the cascaded multibit modulator much more tolerant of DAC nonlinearity than the single-stage modulator. To achieve 12-bit dynamic range, the cascaded modulator requires a DAC linearity of only 8 bits, while more than 10-bit linearity is needed in the single-stage modulator. Moreover, even with 10-bit DAC linearity, the single-stage modulator still exhibits harmonic distortion. In contrast, the harmonic power produced by the cascaded modulator is well below the power of the baseband quantization noise.

4. System Simulation

In the design of an analog $\Delta\Sigma$ modulator, simulation is necessary. Analytical models can provide good starting points in the design of an analog $\Delta\Sigma$ modulator, but simulation is needed to determine the effects of analog nonideality on system performance. In this work, MATLAB is used to simulate the system performance and the requirement of the building blocks in the modulator.

In MATLAB, the whole structure of the programmable $\Delta\Sigma$ is established. The performance of the modulator ADC can be simulated by adjusting the parameters in the SIMULINK environment. The following three figures (Figure 5.5, Figure 5.6 and Figure 5.7) show the function of the $\Delta\Sigma$ modulator ADC under the worst blocker conditions in DCS1800 mode. Table 5.2 shows the final output SNR of the modulator A/D converter with the programmable decimation filter. The achieved SNRs are 23.27dB/28dB/20dB for WCDMA/DECT/DCS1800 respectively, which are enough for the demodulation of the original signal. Note that the decimation filter for DECT and DCS1800 not only filters out the blockers but also completes the channel selection function.

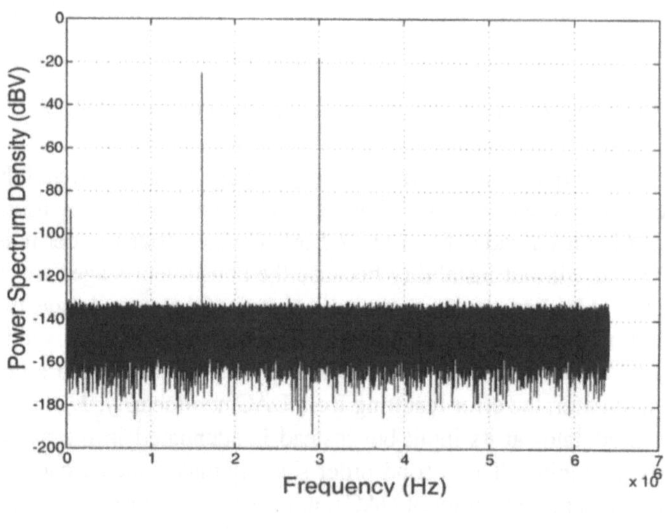

Figure 5.5. **Input Signal to the $\Delta\Sigma$ Modulator with Blockers**

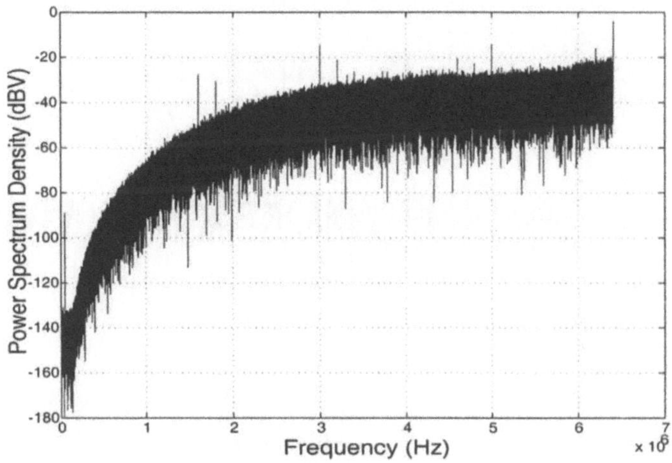

Figure 5.6. Output Spectrum of the Analog ΔΣ Modulator

Table 5.2. Ideal Simulation of the ΔΣ Modulator ADC with Decimation Filter

	WCDMA	DECT	DCS1800
Sampling Frequency	40MHz	22.4MHz	12.8MHz
Oversampling Ratio (OSR)	8	16	64
Decimation	6-order Sinc Filter with one half-band filter	6-order Sinc Filter with two half-band filters	6-order Sinc filter with two half-band filters
SNR w/1bit	6dB	13dB	20dB
SNR w/3bit	23.27dB	27.8dB	20dB

4.1. Effect of the Finite OTA Gain

The finite gain of an integrator introduces simultaneously a gain and a pole error in the integrator transfer function. This can be modeled by equation 5.14

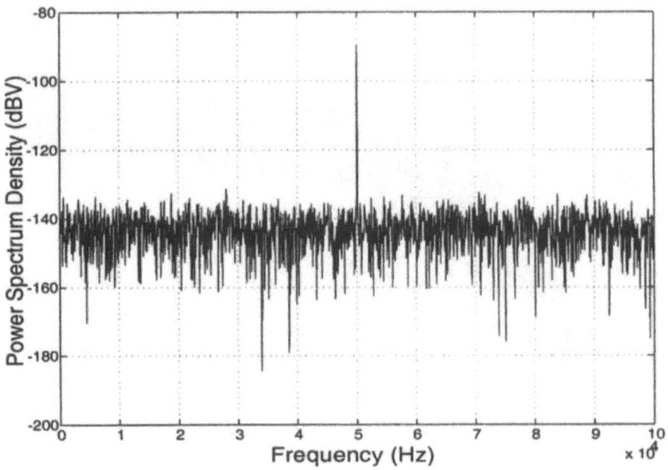

Figure 5.7. Output Spectrum after Decimation Filter

[14] [49].

$$H(z) = \frac{1}{1+1/A} \times \frac{z^{-1}}{1-(1+1/A)z^{-1}} \tag{5.14}$$

where A is the finite OTA dc gain. These errors are especially important in the first and second integrators of the first stage. On the one hand, the gain error limits the perfect cancellation of the first-stage quantization noise. On the other hand, the pole error creates noise leakage at low frequencies that is not cancelled by the digital noise cancellation network and that can degrade significantly the converter performance.

Figure 5.8 shows the finite OTA gain effect on the proposed architecture under DCS1800 and DECT modes. For WCDMA mode, because of the low oversampling ratio, the finite OTA gain results in very small degradation of the performance. Based on the above simulation results, the gain requirement of the four OTAs is given in Table 5.3. Note that for a Nyquist converter, such as pipelined A/D converter, to obtain the 14bit performance of DCS1800 mode, the OTA gain must be at least 90dB. Thus, the advantage of using $\Delta\Sigma$ techniques is to relax the performance of OTAs, thus saving power consumption.

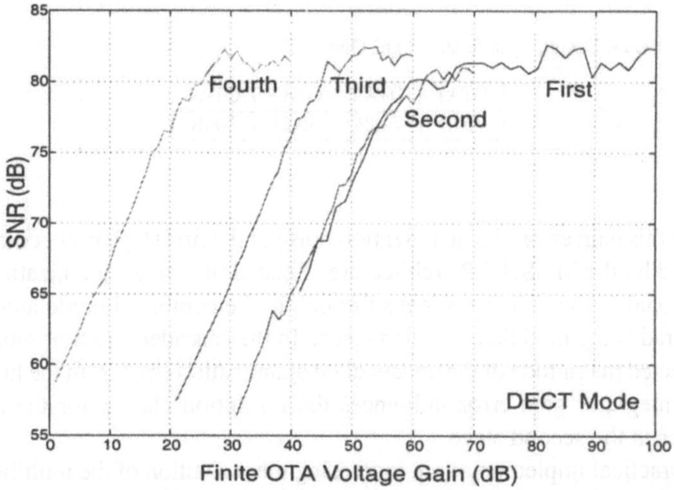

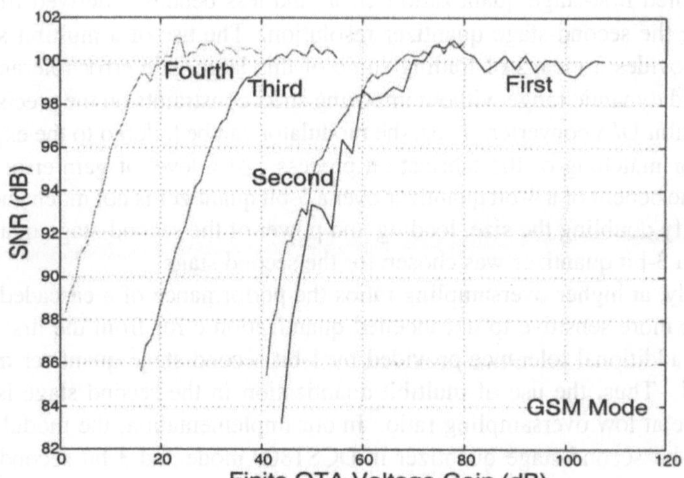

Figure 5.8. OTA DC Gain Effect on Modulator Performance

4.2. Gain Mismatch or Gain Error

Mismatch between the sampling and integrating capacitors results in a gain error in the transfer function of the switched-capacitor integrator [17]. As a

Table 5.3. OTA Design Specification: DC Gain

OTA1	OTA2	OTA3	OTA4
70dB	70dB	50dB	30dB

result of this gain error, the quantization noise of the first stage is not completely cancelled by the MASH 2-2 architecture. Shaped first-stage quantization noise proportional to the gain error leaks through to the combined modulator output and degrades the modulator performance. In the cascaded multibit modulator, the expected magnitude of the uncancelled quantization error from the first stage due to integrator gain error influences the resolution chosen for the multibit quantizer in the second stage.

In a practical implementation, increasing the resolution of the multibit quantizer reduces the second-stage quantization error and thereby increases the sensitivity to uncancelled quantization error from the first stage. With increased levels of gain error, the performance of the modulator becomes dominated by uncancelled first-stage quantization error, and less benefit is derived from increasing the second-stage quantizer resolution. The use of a multibit second stage provides a means of trading some of this large gain error tolerance for increased dynamic range without imposing strict constraints on the precision of the multibit D/A converter. Thus, the modulator can be tailored to the expected capacitor matching of the fabrication process. At a level of gain error about 0.2%, the benefit of a 4-bit quantizer over a 3-bit quantizer is not much and does not justify doubling the size, loading and power of the second-stage quantizer. Hence, a 3-bit quantizer was chosen for the second stage.

Finally, at higher oversampling ratios the performance of a cascaded modulator is more sensitive to uncancelled quantization error from the first stage, and the additional tolerance provided by 1-bit second-stage quantizer may be required. Thus, the use of multibit quantization in the second stage is most attractive at low oversampling ratio. In our implementation, the modulator is set to 1-bit second-stage quantizer in DCS1800 mode and 3-bit second-stage quantizer in DECT and WCDMA modes.

In the system simulation, the coefficients in every integrator is statistically changed using Monte Carlo simulation and the performance is then evaluated by calculating output SQNR. As we discussed, for DCS1800 mode, the performance is limited by the input referred thermal noise or KT/C noise. The uncancelled quantization noise because of the matching problem is not the dominant issue. However for DECT mode, the quantization noise becomes the factor that limits the modulator performance and thus the mismatch between the analog gain and the digital gain should be quantified. Figure 5.9 shows the Monte Carlo simulation of the coefficients mismatch effect and Table 5.4 concludes the results. In Table 5.4, a_1, a_2, a_3 and a_4 refer to the coefficients in the four integrators respectively.

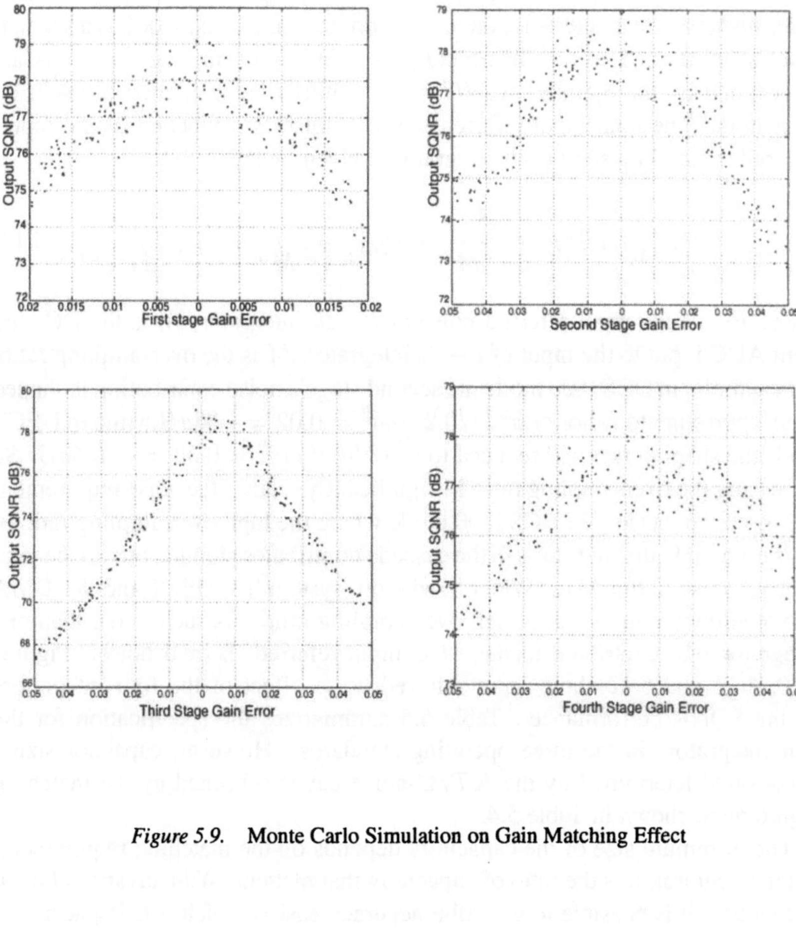

Figure 5.9. Monte Carlo Simulation on Gain Matching Effect

Table 5.4. Integrator Gain Matching Requirement

	a_1	a_2	a_3	a_4
Nominal Value	0.2	0.25	0.2	0.5
Matching Requirement	0.25%	2.5%	0.25%	2.5%

4.3. Noise Shaping and Capacitor Scaling

One strategy to minimize power dissipation is to take advantage of noise-shaping in $\Delta\Sigma$ modulators to scale down the capacitors (and bias currents) in later integrators in the modulator chain. The thermal and flicker noise of the second integrator, when referred to the ADC input, is shaped with a first-order high-pass characteristic; the noise from the third integrator has a second-order shaping, and so on. In general, the minimum sampling capacitors are determined by the required KT/C noise, without being limited by parasitic capacitors. The total input-referred thermal noise as a function of the input-referred thermal noise of each integrator is given by [52]:

$$P_{N,tot} = P_{N1}\frac{1}{M} + P_{N2}\frac{\pi^2}{3{A_2}^2M^3} + P_{N3}\frac{\pi^4}{5{A_3}^2M^5} + P_{N4}\frac{\pi^6}{7{A_4}^2M^7} \quad (5.15)$$

where P_{Ni} is the input-referred noise of $i-th$ integrator; A_i is the DC gain from ADC input to the input of $i-th$ integrator, M is the oversampling ratio. For example, in DCS1800 mode the second stage's noise contribution is shaped by an approximate factor of $\pi^2/3/0.2^2/64^2 = 0.02 = -33dB$ while in DECT mode the shaping factor is reduced to $\pi^2/3/0.2^2/16^2 = 0.32 = -10.5dB$. So increasing the oversampling ratio M significantly relaxes the noise requirement of subsequent stages. In DCS1800 mode where the high oversampling ratio of 64 is used, only the first stage in the cascade contributes significantly to the total thermal noise of the ADC. Wider bandwidth system like DECT and WCDMA do not allow the luxury of a high oversampling ratio. As such, more than one integrator will contribute to the ADC input-referred thermal noise. Figures 5.10, 5.11 and 5.12 show input referred noise effect of the four integrators on the SQNR performance. Table 5.5 summarizes the specification for the four integrators in the three operating standards. However, capacitor sizing is not only determined by the KT/C noise but also limited by the matching requirement shown in Table 5.4.

The minimum size of the capacitors depends on the matching requirement. In the integrator, it is the ratio of capacitors that matters. With advanced layout techniques, it is possible to get 10bit accuracy easily, which is 0.1% accuracy [31]. In our design we keep the minimum capacitor at 0.2pF.

4.4. Settling Error and Slew Rate Limitation

The finite bandwidth of the opamp translates into incomplete linear settling in the time domain when the amplifiers are used in switched-capacitor integrators. This also causes an integrator gain error as shown below.

$$H(z) = \frac{C_s}{C_I} \times \frac{z^{-1}}{1 - z^{-1}} \times (1 - e^{-n_\tau}) \quad (5.16)$$

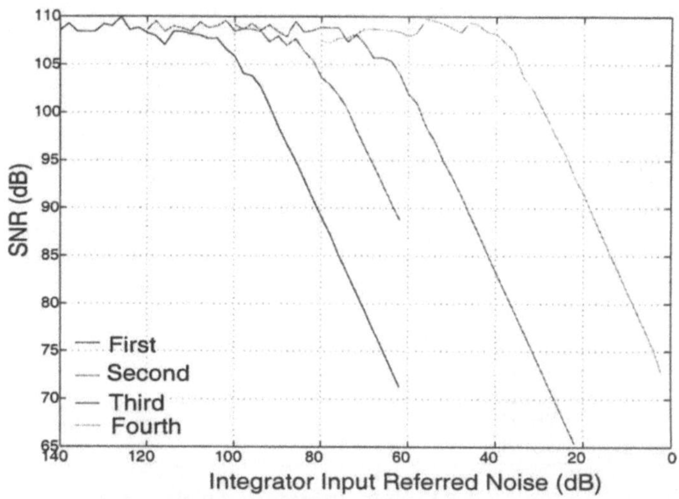

Figure 5.10. Integrator Noise versus Modulator SNR in GSM Mode

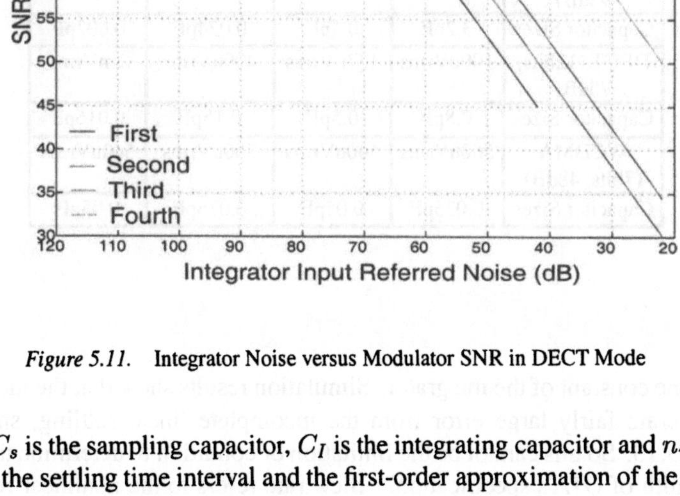

Figure 5.11. Integrator Noise versus Modulator SNR in DECT Mode

where C_s is the sampling capacitor, C_I is the integrating capacitor and n_τ is the ratio of the settling time interval and the first-order approximation of the closed

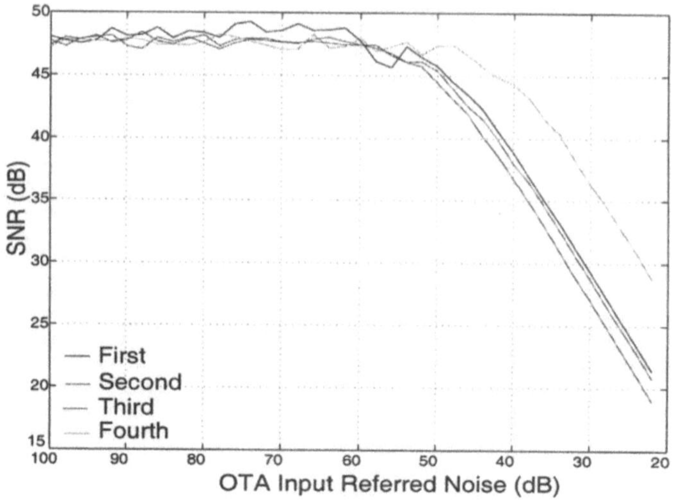

Figure 5.12. Integrator Noise versus Modulator SNR in WCDMA Mode

Table 5.5. Noise Budget and Capacitor Scaling of Integrators

Noise Budget Capacitor Sizing	Integrator I	Integrator II	Integrator III	Integrator IV
GSM (15bits, 95dB)	50uVrms	177uVrms	1mVrms	3mVrms
Capacitor Sizes	3.2pF	0.5pF	0.024pF	0.002pF
DECT (12bits, 75dB)	100uVrms	177uVrms	400uVrms	1mVrms
Capacitor Sizes	0.8pF	0.5pF	0.15pF	0.016pF
WCDMA (7bits, 48dB)	560uVrms	560uVrms	560uVrms	560uVrms
Capacitor Sizes	0.025pF	0.05pF	0.076pF	0.05pF

loop time constant of the integrator. Simulation results show that the modulator can tolerate fairly large error from the incomplete linear settling, shown in Figure 5.13. So this cannot be the limitation or condition to determine the finite bandwidth of OTA's specification. Slew rate refers to the nonlinear response

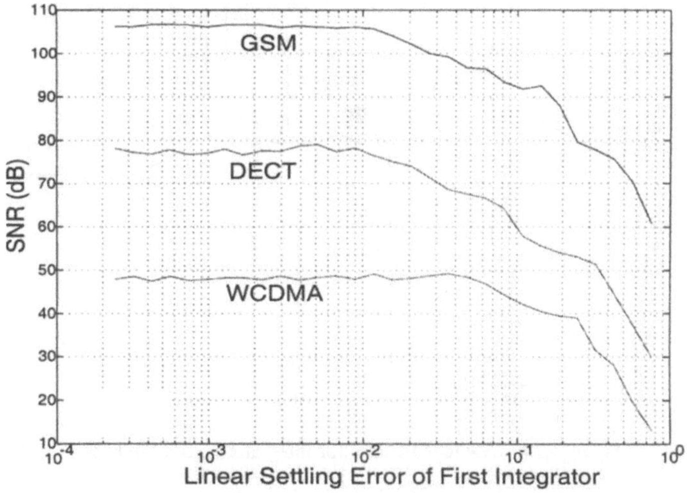

Figure 5.13. Linear Settling Error Effect on Modulator Performance

of an amplifier to a large input step. This nonlinear response occurs because the amplifier can only deliver a maximum amount of differential current due to its class A biasing. As a result, the amplifier output cannot track the large rate of change at the amplifier input. To model and simulate the slewing and partial settling is extremely complex due to the nonlinear dynamics of the integrator as well as the $\Delta\Sigma$ modulator [48].

One rough way to determine slew rate and the settling time constant is by separating the settling process into two parts. The percentage of slewing in the integrating phase is chosen as $1/4$ of the total settling time and then based on the reference voltage level and the modulator resolution, the required slew rate and settling time constant can be easily calculated. Unfortunately, this method always results in a very large slew rate and a very small settling time constant. This leads to a power-hungry OTA design. Another approach is to use a single pole OTA model and constant g_m assumption, which leads to the impractical conclusion that if the slew rate is large enough, an OTA with a large settling time does not cause severe SNR degradation [59]. To solve this problem, a more accurate modeling of the nonlinear settling behavior of the integrator is presented [71].

4.5. Accurate Slew Rate Modeling

Let us consider the integrator used in a modulator as shown in Figure 5.14.

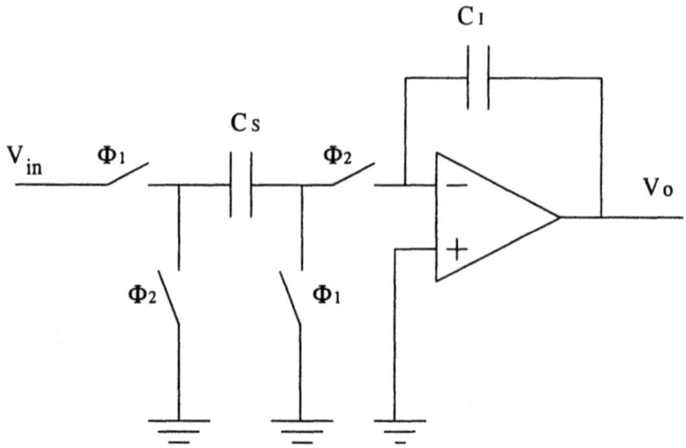

Figure 5.14. Switched-Capacitor Integrator in ΔΣ Modulator

When Φ1 is on and Φ2 is off, the input signal charges C_S. When Φ1 is off and Φ2 is on, the charge on C_S transfers to C_I. Thus the output voltage in the time domain is,

$$v_o(t) = v_o(nT - T) + \frac{C_s}{C_i} V_{in}(nT - \frac{T}{2})(1 - e^{-(t-nT-\frac{T}{2})/\tau}) \tag{5.17}$$

where T is the sampling period, $nT - \frac{T}{2} < t < nT$, τ is the settling time constant of the OTA and is assumed to be a constant. The maximum slope of the settling happens at the beginning when $t = nT - T/2$, which is,

$$\frac{dv_o(t)}{dt}|_{max} = \frac{dv_o(t)}{dt}|_{t=nT-T/2} = \frac{C_s}{C_i}\frac{V_{in}}{\tau} \tag{5.18}$$

If the slew rate (SR) of the OTA is larger than the above value, no slewing happens and 5.17 is enough to describe the whole settling behavior. When SR is smaller than the above value, $v_o(t)$ increases or decreases according to the sign of the input signal with the slope of SR or $-SR$ until the linear settling slope of OTA is smaller than SR. The closed-form equations of this process are derived [51] according to the above analysis. The following equations express the model:

$$t_0 = \frac{\frac{C_s}{C_i}|V_{in}|}{SR} - \tau \tag{5.19}$$

$$y = sgn(V_{in}) \times SR \times t_0; \tag{5.20}$$

$$v_o(nT) = \frac{C_s}{C_i} V_{in}(1 - ye^{\frac{(t_0-\frac{T}{2})}{\tau}}) \tag{5.21}$$

In a small-signal analysis of the folded-cascode OTA, g_m is considered to be constant.

With first-order approximation, the slew rate SR and the unit-gain frequency ω_t are found to be:

$$SR = \frac{I_{tail}}{C_L} \tag{5.22}$$

$$\omega_t = \frac{g_m}{C_L} \tag{5.23}$$

where I_{tail} is the tail current, C_L is the load capacitor. However, in switched-capacitor ΔΣ modulators, the input of the OTA operates in a large signal range. The nonlinearity of g_m must be included in the simulation for high-resolution modulators. In a large-signal analysis of a differential common source pair, the relationship between the output differential current and the input differential voltage is [24].

$$I_d = \frac{1}{2}\mu C_{ox}\frac{W}{2L}V_{id}\sqrt{\frac{2I_{ss}}{\mu C_{ox}W/2L} - (V_{id})^2} \tag{5.24}$$

Where I_d is the output differential current, V_{id} is the input differential voltage and I_{ss} is the tail current. The transconductance g_m can be calculated by differentiating equation 5.24.

$$g_m = \frac{dI_d}{dV_{id}} = \eta g_{m0} \tag{5.25}$$

$$\eta = \sqrt{1 - \alpha V_{id}^2} - \frac{\alpha V_{id}^2}{\sqrt{1 - \alpha V_{id}^2}} \tag{5.26}$$

where $\alpha = \frac{\mu_n C_{ox} W/L}{4I_{ss}}$ and $g_{m0} = \sqrt{\mu_n C_{ox} I_{ss} W/L}$. The large-signal settling time constant can be expressed as:

$$\tau = \frac{C_L}{\beta g_m} = \frac{1}{\eta}\frac{C_L}{\beta g_{m0}} = \frac{1}{\eta}\tau_0 \tag{5.27}$$

where β is the closed loop feedback coefficient of the integrator. Here we see that the settling time constant is not only determined by the size and biasing of the transistors but also a function of the input differential voltage. Since this nonlinearity is signal-dependant, harmonics will be produced. This effect is critical to the design, especially for high resolution ΔΣ modulator. It has been shown that nonlinear settling time constant τ also contributes to the total nonlinear distortion of the settling. One way to consider this effect is to establish the nonlinear differential equation and solve it in closed-form. This is very complex if not impossible. In our work, a simple algorithm is used. Note that the relationship between the differential input and output voltage of the OTA in Figure 5.14 is found to be [82]:

$$V_{id}(t) = \frac{V_o(t)}{A} - \beta\tau_0\frac{dV_o(t)}{dt} \tag{5.28}$$

where A is the DC gain of the OTA. With (5.27) the obtained result on nonlinear τ can be easily included in the simulation.

The algorithm is described as follows:

1 The total settling time T is divided into N steps. Every step has a time of T/N;

2 Compare the maximum settling slope using (5.18). If SR is larger, go to step 4, otherwise go to step 3;

3 Increase (or decrease) V_o with the slope of SR (or $-SR$). Calculate new settling slope in (5.18) with the changed input V_{in}. Go to step 2;

4 Calculate V_{id} using 5.28. Calculate τ using equation 5.26 and 5.27. Calculate the next output voltage V_o using the following equation:

$$V_o(t) = V_o(t - \frac{T}{N}) + V_{in}(e^{-(t-\frac{T}{N})} - e^{-t}) \tag{5.29}$$

Repeat step 4 until $t = T$.

As an example, the fourth-order 2-2 cascade $\Delta - \Sigma$ modulator is simulated in the GSM mode and the results are compared with other approaches. The simulated SNR is 93dB for signal bandwidth of 100KHz, after considering all the nonideal effects.

Figure 5.15 shows the simulation results for the slew and non-linear settling. The lower figure is the power spectrum of the modulator output with the SR of 75V/us and τ_0 of 4ns, which shows that there is no visible harmonic tones. The upper figure gives the power spectrum with the SR of 65V/uS and τ_0 of 4ns. In this case, the harmonic tones can be obviously seen and is about 50dB above noise floor which leads to THD of about -70dB, severely degrading the SNDR of the whole $\Delta\Sigma$ modulator. Therefore the conclusion can be drawn that a SR of 75V/us and τ_0 of 4ns is needed for the first OTA.

The simulation results using the other two modeling approaches, described previously, are also compared here. Table 5.6 lists the simulation results obtained from these three approaches. It can be observed that method 1 (linear settling assumption) suggests that a high resolution $\Delta\Sigma$ modulator can be constructed with low bandwidth and high slew rate integrators. In practice, it is difficult to achieve high-resolution data conversion using low bandwidth integrators [59]. While method 2 (rough estimation) always overestimates the SR

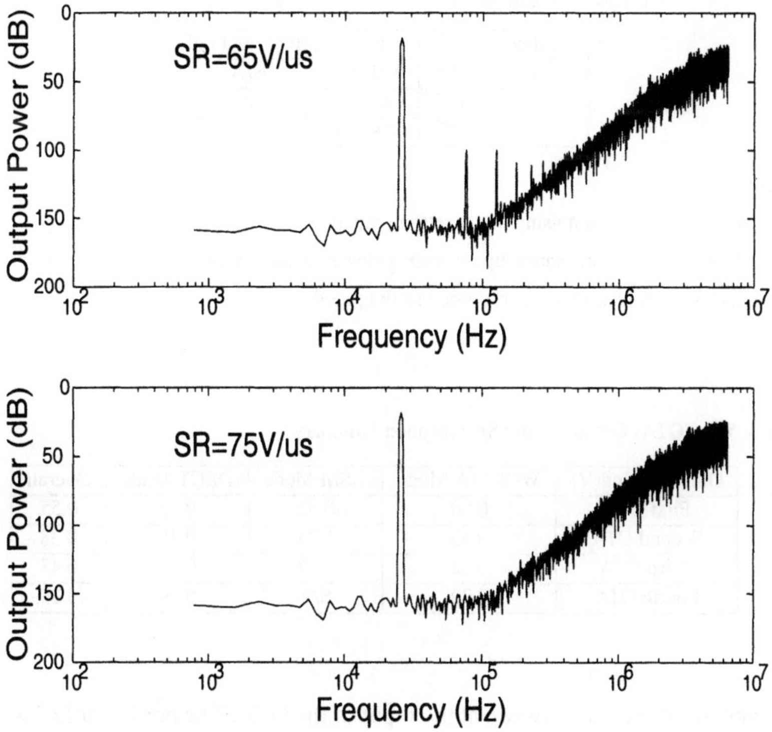

Figure 5.15. **Power spectrum of modulator output with SR=75V/us and SR=65V/us**

and τ_0 requirement on the OTA. The proposed method gives us a more accurate and practical estimation. In this example, the SR and τ_0 requirement can be relaxed by one half, which results in at least 50% power consumption saving. Therefore, it provides a better guidance to optimize the circuit.

4.6. Opamp Output Swing

To avoid nonlinearity, the output swing capability of each OTA should be large enough so that the largest signal at the output node can be processed without limiting or clipping. This can be easily simulated using a limiter model in MATLAB. The results are shown in the Table 5.7. Though we discussed the DC gain requirement in terms of gain error effect, nonlinearity error from the signal-dependent DC gain in real OTA should also be taken into account. The typical OTA's DC gain in terms of output voltage is shown in Figure 5.16. In this

Table 5.6. OTA Slew Rate Estimation

Methods	SR	τ_0	Tail current ratio
1	20	100ns	N/A
2	150	2ns	2
3	75	4ns	1

Note:

- Method 1–simulation using linear settling model
- Method 2–rough estimation by separating slew rate and settling
- Method 3–simulation using proposed settling model

Table 5.7. OTAs Output Swing Specification Summary

Output Swing(V)	WCDMA Mode	GSM Mode	DECT Mode	Overall
First OTA	0.50	0.52	0.55	0.55
Second OTA	0.35	0.38	0.38	0.38
Third OTA	0.22	0.47	0.25	0.47
Fourth OTA	0.39	N/A	0.39	0.39

Figure, we did the DC sweep at the input of the OTA. The nonlinear DC gain in Figure 5.16 is obtained by calculating the derivative of the output signal. It is shown that with the output swing of 3.0Vpp (differential), shown in the bottom part of the figure, the OTA DC gain is in the range from 80dB to 85dB, shown in the top part of the figure. The overall nonlinearity in the integrator introduced by the DC gain should be less than the resolution required. Since the OTA works in a feedback loop, the nonlinearity can be greatly reduced as long as the OTA's DC gain is large enough [24]. According to our simulation, for 14bit(86dB) resolution, a 70dB OTA gain with $\pm 2.5dB$ gain variation degrades the SNDR performance by 1.7dB while a 80dB OTA gain with $\pm 2.5dB$ gain variation introduces only 0.4dB SNDR degradation. So the OTA DC gain specification is defined by the nonlinearity requirement rather than the gain error requirement.

4.7. Multi-bit DAC Nonlinearity

Comparator hysteresis is not critical for $\Delta\Sigma$ modulator [14]. According to our simulation, shown in Figure 5.17, 40mV hysteresis is small enough so that no performance degradation happens.

As discussed in section 2, the required nonlinearity of the internal multi-bit DAC can be relaxed by placing it in the second stage. For the required

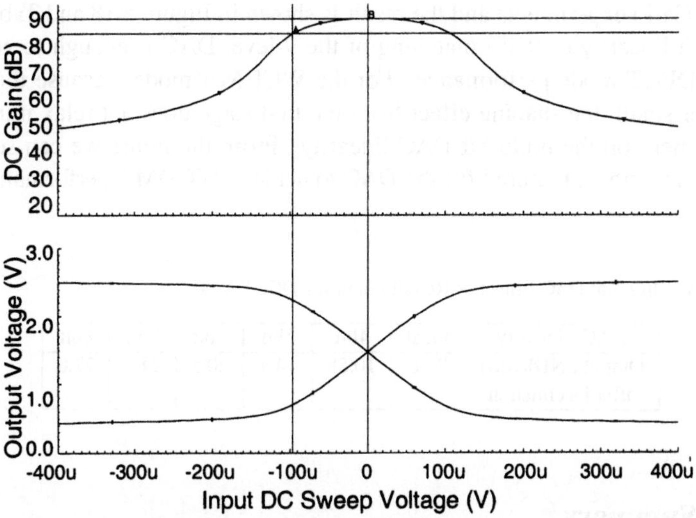

Figure 5.16. **OTA DC Gain versus Output Swing**

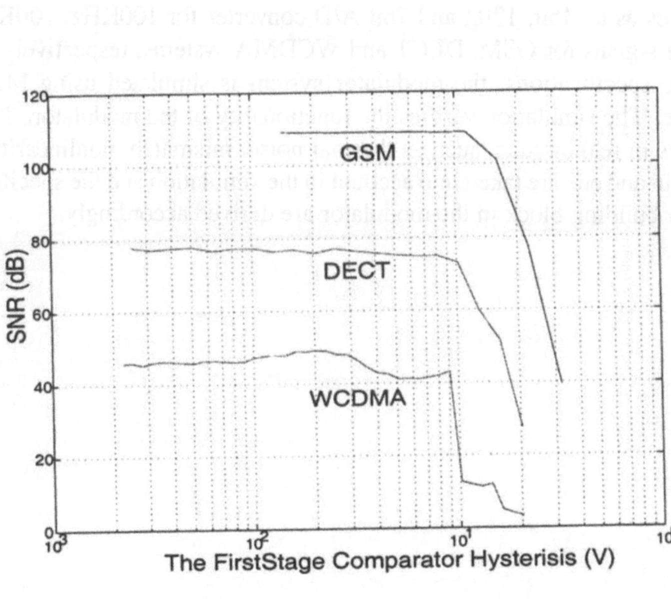

Figure 5.17. **Comparator Hysteresis on Modulator Performance**

12bit DECT mode, the 9-level DAC's linearity specification is simulated using Monte Carlo experiments and the result is shown in Figure 5.18 and Table 5.8. An 8-bit linearity or 0.4% matching of the 9-level DAC is enough to achieve 12-bit DECT mode performance. For the WCDMA mode, because the OSR is rather small, the shaping effect from the first stage does not relax much the requirement on the multi-bit DAC linearity. From the figure we can see that 6-7 bit linearity is required for the DAC to achieve WCDMA performance.

Table 5.8. Internal DAC Linearity Requirement for DECT Mode

DAC Linearity	Ideal	4bit	5bit	6bit	7bit	8bit
Output SNDR(dB) after Decimation	27.8	10.88	14.7	20.5	24.8	27.0

5. Summary

The programmable $\Delta\Sigma$ modulator is one of the major building blocks in the multi-standard receiver. By adjusting the OSR and reconfiguring the circuit, it behaves as a 14bit, 12bit and 7bit A/D converter for 100KHz, 700KHz and 2.5MHz signals for GSM, DECT and WCDMA systems, respectively. Based on these specifications, the modulator system is simulated using MATLAB software. The simulation verifies the functionality of the modulator. The non-idealities in real circuits, such as thermal noise, mismatch, nonlinearity, finite OTA gain and etc, are taken into account in the simulation and the specifications of every building block in the modulator are derived accordingly.

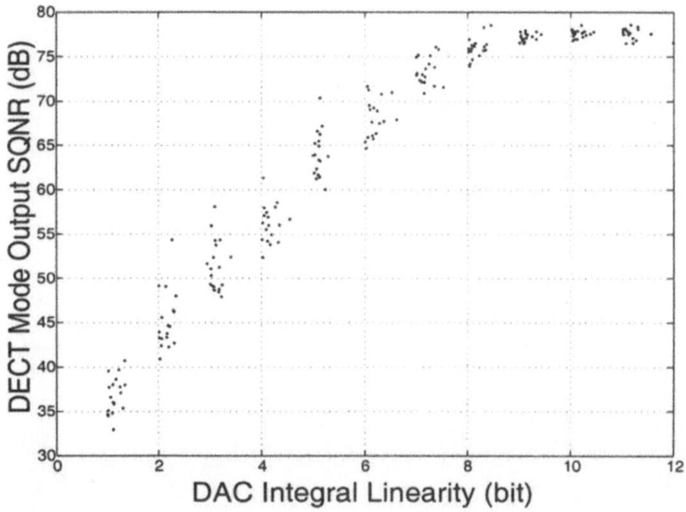

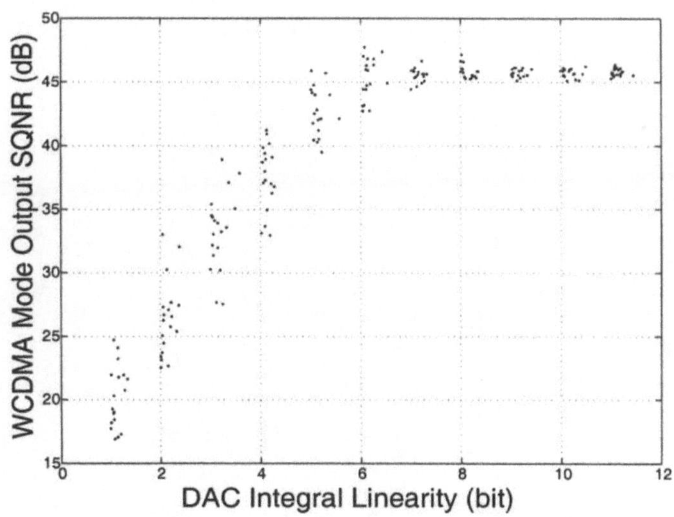

Figure 5.18. 9-Level DAC Linearity Requirement for DECT and WCDMA Mode

Chapter 6

MULTI-STANDARD ΔΣ MODULATOR: DESIGN

This chapter presents the design of building blocks in the programmable switched-capacitor ΔΣ modulator based on the analysis and simulation presented in Chapter 5. A fast simulation method, at the device level, for ΔΣ modulators is also discussed.

1. Switched-Capacitor Integrator

The modulator includes four switched-capacitor integrators. Figure 6.1 shows the first integrator used in the modulator. The integrator is implemented in a fully-differential configuration and employs a two-phase non-overlapped clock set. The input is sampled during phase 1 (Φ_1 and Φ_{1d}) and the charge is transferred from the sampling capacitor (C_S) to the integrating capacitor (C_I) during phase 2 (Φ_2 and Φ_{2d}). At the same time, the appropriate 1-bit DAC reference level, controlled by the following quantizer output (Q), is applied at the bottom plate of C_S. The integrator employs the bottom-plate sampling technique to minimize signal dependent charge-injection. This is achieved by delayed clocks (Φ_{1d}, Φ_{2d}) [34]. When switches labeled Φ_1 are first turned off, the charge injection from these switches, to the first order, is independent of the input signal since their source and drain nodes are always connected to the common mode (analog ground level). Then because the top plate is now floating, turning off Φ_{1d} switches does not introduce charge-injection errors. Note that the feedforward and feedback gains of the first integrator are 0.2 and -0.2 respectively. The implementation of Figure 6.1 allows the sharing of the sampling capacitor for the two paths. There are two advantages of doing this. One is that $\frac{KT}{C}$ noise is reduced by half compared with using two sampling capacitors. The other is that it minimizes the loading on the OTA and maximizes the feedback factor. Both these two points help reduce the OTA's power consumption.

There are two sets of sampling and integrating capacitors: C_{s1}, C_{I1} and C_{s2}, C_{I2}. In DCS1800 mode, the standard switch "GSMEnable" is on, thus both of the two sets of capacitors are connected and included in the integration process. In WCDMA/DECT mode, GSMEnable is off, thus only C_{s1} and C_{I1} are included in the integration operation.

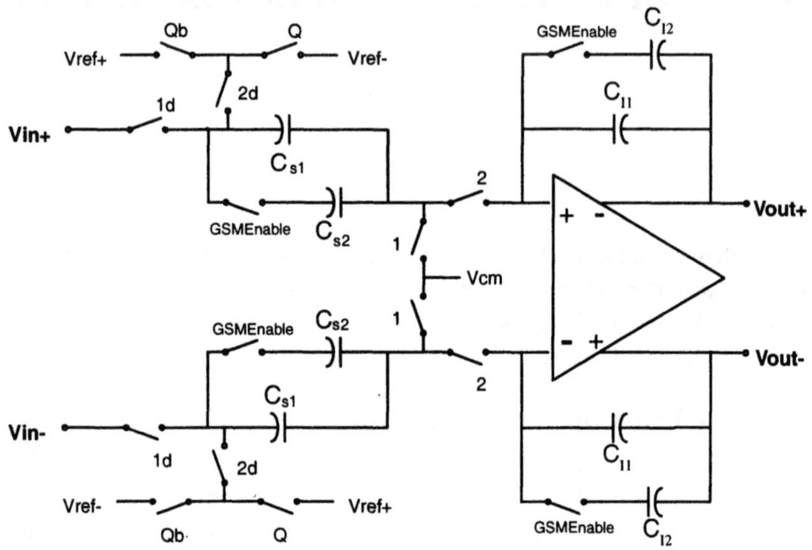

Figure 6.1. Switched-Capacitor Implementation of First Integrator

2. MOS Switch

MOS transistors are used to implement analog switches in the modulator. However the on-resistance between the source and drain is not constant when the input signal varies. In $\Delta\Sigma$ modulators, a specific sample-and-hold circuit is not required. The input switch actually acts as the sampling switch of the whole modulator A/D. Thus its linearity is an important factor that influences the linearity of the modulator. It is desirable to have a constant on-resistance that is independent of both V_{ds} and V_{gs}. For a square-law NMOS device that operates in the linear (triode) region, the resistance can be expressed as [60]:

$$R_{on} = \frac{1}{\mu_n C_{ox} \frac{W}{L} (V_{GS} - V_{TH})} \tag{6.1}$$

where μ_n is the electron mobility in the channel, C_{ox} is the gate oxide capacitance per unit area, W and L are the effective width and length of the device, respectively, V_{GS} is the gate-source voltage, and V_{TH} is the threshold voltage. The nonlinearity of R_{on} comes from several factors:

- V_{DS} and V_{GS} are signal dependent and R_{on} is a function of V_{DS} and V_{GS}.
- Since the source and the back-gate cannot be connect together, the body effect affects the threshold voltage of the MOS switch and R_{on} is also a function of V_{TH}.
- Power supply drifting or ground bounce directly influence V_{GS}.

Another point is that harmonic distortion from the nonlinear switch is also frequency-dependent. This can be easily explained by investigating the transfer function of the equivalent circuit of the sampling circuit in Figure 6.2, which is given by,

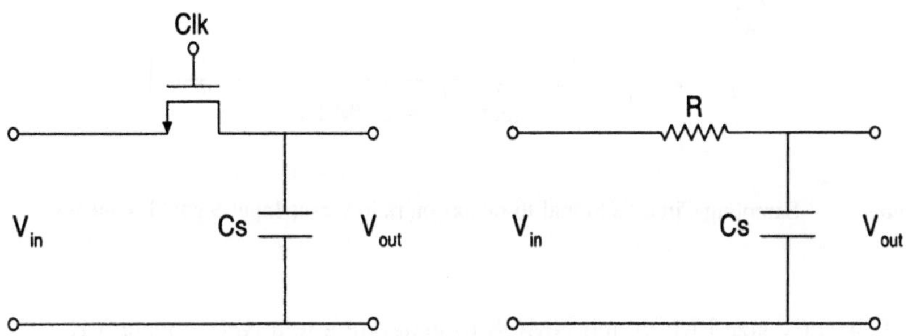

Figure 6.2. MOS Sampling Circuit and its corresponding equivalent circuit

$$H(j\omega) = \frac{1}{1 + j\omega RC_s} \tag{6.2}$$

With the increase of the input frequency, the contribution of the term ωRC_s is larger. While remembering that the resistance R is not a constant and is nonlinear, the nonlinearity of the $H(j\omega)$ is higher with higher frequencies. The SPICE simulation of a CMOS switch's performance is shown in Figure 6.3. The figure clearly demonstrates that the CMOS switch's nonlinearity is a basic performance limitation for high speed sampling circuitry. There are several techniques to reduce the harmonic distortion from CMOS switch:

- Reduce the sampling capacitor. But the sampling capacitor is fixed by the $\frac{KT}{C}$ noise requirement.
- Reduce the resistance R and make the "quiescent bandwidth" larger. This can be done by increasing the size of the CMOS transistors. But the switch may become extremely large. Then some other problems are introduced. The larger depletion capacitances can couple more clock-through noise. The driving ability of the clock generator has to be increased too.
- Boost of VDD to reduce R. Then the gate oxide reliability needs to be taken care of.

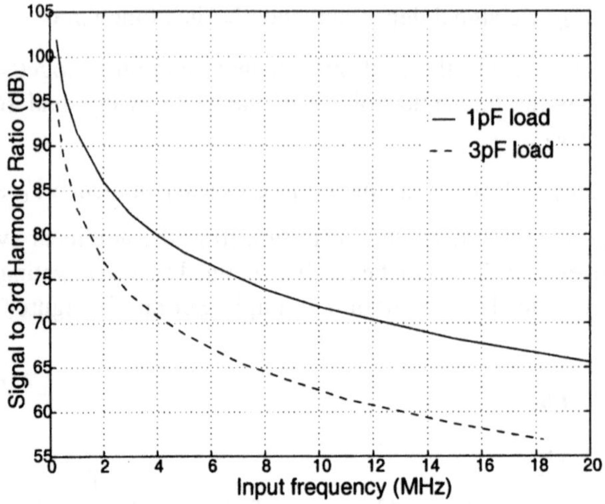

Figure 6.3. Sampling Circuit's Signal-to-distortion ratio versus Input Signal Frequency

- Correctly size PMOS and NMOS to generate a relatively constant R.
- Design extra circuitry to get a constant gate overdrive voltage, which is not signal dependent.

In our implementation, the PMOS and NMOS transistors sizes are adjusted to achieve the best linearity. The on-resistance curve is shown in Figure 6.4. Another important issue in sampling circuits in our application is that blockers may generate intermodulation or harmonic products. Because the blockers locate at rather high frequency range, according to Figure 6.3, which shows the sampling circuit's SDR (signal-to-distortion ratio) versus input signal frequency, the harmonic products could be very high. These products may alias close to the baseband signal, thus putting more filtering requirement on the decimation filter. The simulation in Figure 6.5 shows that these products are well below the signal level (30dB). Though the aliased third harmonics are comparatively larger, these spurs locate at higher frequency than the blockers after aliasing. The decimation filter has even more attenuation on these spurs. So no extra burden on the decimation filter is added from this intermodulation effort.

3. Folded-cascode OTA and Power Optimization

Based on the system level simulation of the $\Delta\Sigma$ modulator, the fully-different amplifier to be used in the first stage of the modulator should have the specifi cations in Table 6.1.

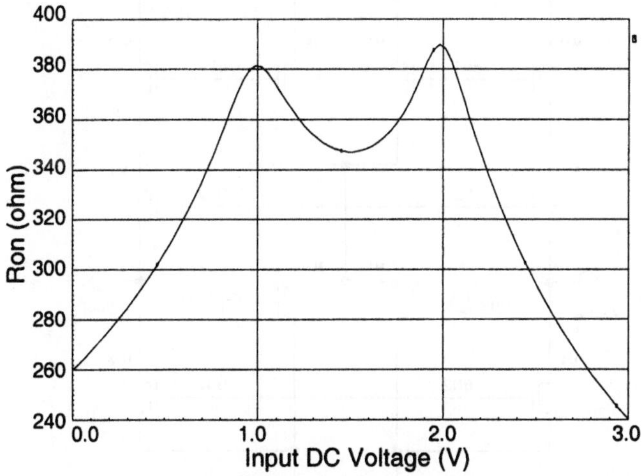

Figure 6.4. On-resistance of the Input Sampling Switch

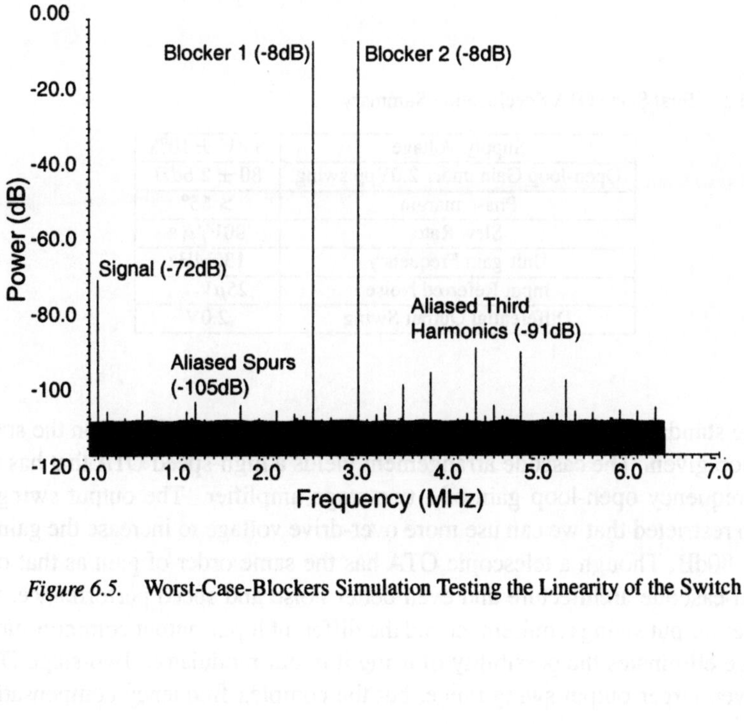

Figure 6.5. Worst-Case-Blockers Simulation Testing the Linearity of the Switch

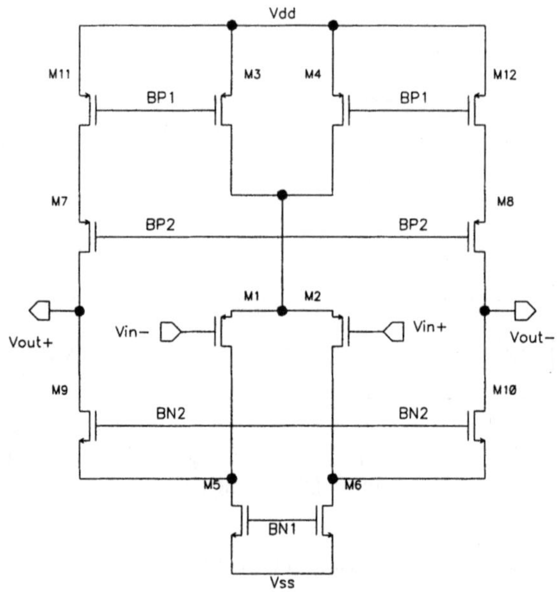

Figure 6.6. Folded-cascode OTA Schematic

Table 6.1. First Stage OTA Specification Summary

Supply Voltage	$3.3V \pm 10\%$
Open-loop Gain under 2.0Vpp swing	$80 \pm 2.5dB$
Phase margin	$> 55^o$
Slew Rate	$90V/\mu$ s
Unit-gain Frequency	130MHz
Input Referred Noise	$25\mu V_{rms}$
Differential Output Swing	2.0V

The standard folded-cascode OTA architecture is chosen based on the specification given. The cascode arrangement yields a high-speed OTA that has the low-frequency open-loop gain of a two-stage amplifier. The output swing is not so restricted that we can use more over-drive voltage to increase the gain to about 80dB. Though a telescopic OTA has the same order of gain as that of a folded-cascode architecture and even better noise and speed performance, the smaller output swing requirement and the different input-output common-mode voltage eliminates the possibility of using it in our modulator. Two-stage OTA has even larger output swing range, but the complex frequency compensation scheme makes the design less robust. Also because the capacitive loads of the OTA are different in three modes, the compensation needs more care in the design. Moreover, it is easy to loose phase margin considering the processing variations. The schematic of the folded-cascode OTA is shown is Figure 6.6. The input differential pair is chosen to be PMOS transistors. The major consideration is the noise performance. Since resolution limitation in the DCS1800 mode is the $1/f$ noise of the first integrator stage, PMOS transistors are chosen since they have less $1/f$ noise than NMOS transistors. Having a PMOS input stage minimizes the output noise due to the $1/f$ noise.

3.1. Input Referred Noise

The input-referred thermal noise power of a single common-source transistor is given by:

$$V_{n,in}{}^2 = 4KT \cdot \frac{2}{3} \cdot \frac{1}{g_m} \cdot \Delta f \tag{6.3}$$

The thermal noise behavior of the differential folded-cascode OTA of Figure 6.6 can be calculated as:

$$V_{nt,in}{}^2 = 8KT(\frac{2}{3g_{m1,2}} + \frac{2g_{m5,6}}{3g^2{}_{m1,2}} + \frac{2g_{m11,12}}{3g^2{}_{m1,2}}) = \frac{16}{3}KT\frac{1}{g_{m1,2}} \cdot \beta \tag{6.4}$$

where $\beta = (1 + \frac{g_{m5,6}}{g_{m1,2}} + \frac{g_{m11,12}}{g_{m1,2}})$. At low frequency, the noise of the cascode devices M7, M8, M9 and M10 is negligible. When the amplifier is placed in

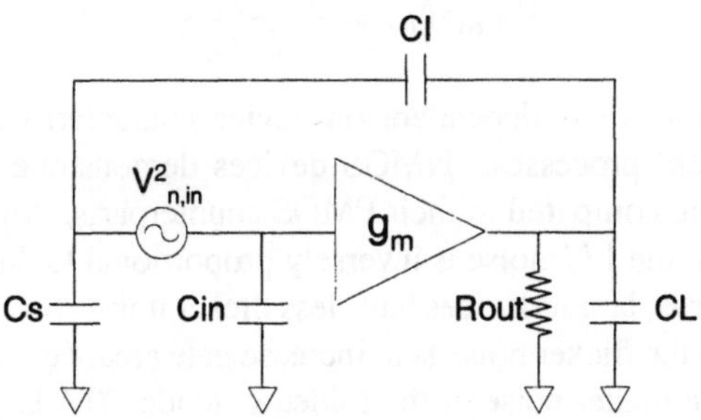

Figure 6.7. OTA Noise Model in the Integrator

the switched-capacitor integrator, as shown in Figure 6.7, the total noise power is evaluated by integrating the noise spectrum to infinity [62]. This is because the sampling process causes the noise spectrum at high frequency to fold down to baseband. The thermal noise of the OTA, when referred to the output of the integrator, as in Figure 6.7, is given by:

$$V_{n,o}{}^2 = \frac{4}{3} \cdot \beta \cdot \frac{KT}{fC} \tag{6.5}$$

$$C = C_L + \frac{C_I(C_{in} + C_s)}{C_I + C_{in} + C_s} \tag{6.6}$$

$$f = \frac{C_I}{C_I + C_{in} + C_s} \tag{6.7}$$

Where C is the total load capacitor of the OTA and f is the feedback factor. Then, the input referred fully differential noise $V_{n,in}$ is given by dividing the output noise in the above equation by the square of the closed loop gain.

$$V_{n,in}{}^2 = \frac{4}{3} \cdot \beta \cdot \frac{KT}{fC} \cdot \frac{1}{A_C{}^2} \approx \frac{4}{3} \cdot \beta \cdot \frac{fKT}{C} \tag{6.8}$$

So the thermal noise contribution of the amplifier is also inversely proportional to a capacitance, in this case the total load capacitance. To decrease the factor β, it is better to increase $g_{m1,2}$ and decrease $g_{m5,6}$ and $g_{m11,12}$.

3.2. Flicker Noise

Flicker noise is due to the random trapping and detrapping of minority carriers in the channel of MOS devices [34]. Because the fluctuation in the channel charge carrier density has a relatively large time constant, the noise power density is inversely proportional to frequency. The flicker noise in MOS transistors is modeled as a voltage source in series with the gate and is given by:

$$V_{fl,in}{}^2(f) = \frac{K_f}{WLC_{ox}f} \tag{6.9}$$

where the constant K_f is dependent on device characteristics and can vary widely for different processes. NMOS devices demonstrate a higher flicker noise characteristic compared to their PMOS counterparts. An important point to note here is that the $1/f$ noise is inversely proportional to the transistor area, WL. In other words, larger devices have less flicker noise. So a straightforward strategy to reduce the flicker noise is to increase gate area, i.e. width and length, of the device. The flicker noise of the folded-cascode OTA is given by:

$$V_{nf,in}{}^2 = \frac{2K_P}{(WL)_{1,2}C_{ox}f} + \frac{2K_N}{(WL)_{5,6}C_{ox}f} \frac{2g_{m5,6}}{3g^2{}_{m1,2}} + \frac{2K_P}{(WL)_{11,12}C_{ox}f} \frac{2g_{m11,12}}{3g^2{}_{m1,2}} \tag{6.10}$$

where K_P and K_N is the flicker noise coefficient. Note that PMOS has a much smaller flicker noise coefficient, the input transistors are chosen as PMOS in the OTA. The corner frequency is where the flicker noise is equal to the thermal noise. In the implementation process, the corner frequency is around 10KHz to 100KHz, which means that for DCS1800 mode the dominant noise is flicker noise. For design simplicity, the OTA flicker noise contribution is reduced by increasing the sizes of the input and load devices. By increasing both the width and length by the same factor, the gate area is increased without changing the W/L ratio. For a fixed current, this keeps a constant transconductance (g_m) and saturation voltage $V_{DS}{}^{sat}$. Increasing the input device sizes, however, increase the input parasitic capacitances, which in turn degrades the feedback factor. This is tolerable in the modulator because of the relatively large sampling and integrating capacitors.

3.3. DC Gain and Gain Nonlinearity

The low-frequency gain of the amplifier can be expressed as:

$$A_{DC} = g_{m1} \cdot [g_{m9} \cdot r_{o9} \cdot (r_{o5}//r_{o1})]//[g_{m7} \cdot r_{o7} \cdot r_{o11}] \tag{6.11}$$

The transistor output resistance, thus the DC gain, is proportional to the channel length of the device. Since M5, M6, M11 and M12 do not capacitively load the signal path, they can be made to have a long channel to maximize r_{o5} and r_{o11} in the DC gain equation. The PMOS and NMOS legs should be adjusted such that the parallel output impedances are approximately equal. The overall gain is calculated to be 81.5dB. The gain variation over the full swing range is about $\pm 2.5dB$.

3.4. Unity-gain Bandwidth and Slew Rate

The folded-cascode OTA's unity gain frequency is:

$$\omega_u = \frac{g_{m1,2}}{C_L} \tag{6.12}$$

when the OTA is put into the closed loop of the integrator, the closed-loop unity-gain bandwidth is multiplied by the feedback factor. That is,

$$\omega_{uc} = \frac{g_{m1,2}}{C_L} \times f \tag{6.13}$$

Here we see that the sizes of input transistors are the main factor that influences all the specifications. When we increase the size, the flicker noise is decreased and $g_{m1,2}$ is increased, which may lead to higher unity-gain frequency. However, the input capacitor of the OTA also deteriorates the feedback factor, which lowers the OTA's speed. When a large input step is applied to the amplifier, its output is unable to track the large rate of change in its input. This is due to the limit in the differential current which a class A amplifier can deliver. The slew rate of a folded-cascode OTA is determined by the tail current and the load capacitor. That is,

$$SR = \frac{2I_{M11,12}}{C_L} \tag{6.14}$$

The current consumption is determined by the slew rate requirement. Based on the system simulation and some margin, the design target of the slew rate is set to be 125V/us, which leads to a tail current of 500μA with a load capacitance of 4pF in the DCS1800 mode. For DECT and WCDMA modes, the SR requirement is 200V/us with the capacitive load of 1.5pF. So the tail current is 300μA. The unity gain frequency is then satisfied by adjusting the input transistor size. Note that since in DECT/WCDMA modes, the tail current is 40% smaller than that in the DCS1800 mode, the mode switch signal is also used to switch the biasing circuitry to save power in the DECT and WCDMA modes.

3.5. Common-Mode Feedback

Common-mode feedback is required in fully-differential amplifiers to define the common-mode voltages at the high-impedance output nodes. The amplifier uses a switched-capacitor common-mode feedback circuit [34]. In this approach, capacitors C_3 and C_4 generate the average of the output voltages, which is used to create control voltages for the opamp current sources. The dc voltage across C_3 (C_4) is determined by capacitors C_1 (C_2), which are switched between bias voltages and being in parallel with C_3 (C_4). This circuit acts much like a simple switched-capacitor low-pass filter having a dc input signal. The bias voltages are designed to be equal to the difference between the desired common-mode voltage (V_{cm}) and the desired control voltage used for the opamp current sources. In our case, the biasing voltage is set to be BN1 and the control voltage V_{ctl} is expected to be close to BN1 after the common-mode voltage converges. Actually each of transistors M5 and M6 is split into two transistors respectively. One bias voltage is from BN1, the other one is from V_{ctl}. The capacitors being switched, C_1 and C_2, might be between one-fourth and one-tenth the sizes of the non-switched capacitors, C_3 and C_4. Using large capacitance values overloads the opamp more than is necessary during the phase Φ_2, and their size is not critical to circuit performance. Reducing the capacitors too much cause common-mode offset voltages due to charge injection of the switches. Normally, all of the switches would be realized by minimum-size n-channel transistors only, except for the switches connected to the outputs, which might be realized by transmission gates to accommodate a wider signal swing.

4. Timing Sequence and Clock Generators

To minimize signal-dependent charge-injection errors, the integrators need a 2-phase non-overlapping clock with delays. Figure 6.9 shows the timing waveform of the 2-phase clock. Φ_{1d} and Φ_{2d} are the delayed versions of Φ_1 and Φ_2, respectively. The rising edges of the delayed clocks should be lined up with the rising edges of the non-delayed versions in such a way not to decrease

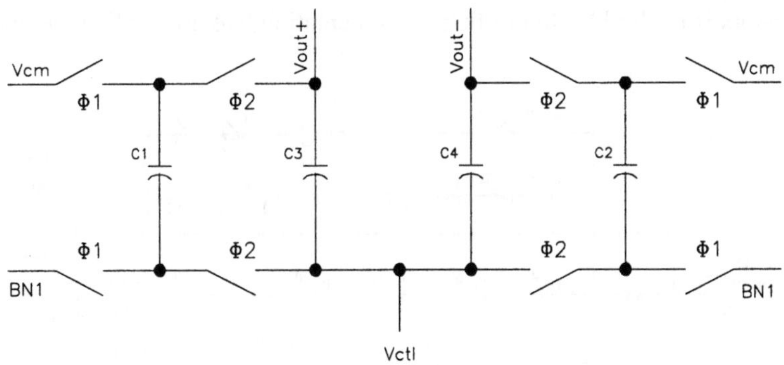

Figure 6.8. Switched-capacitor Common-Mode Feedback Circuit

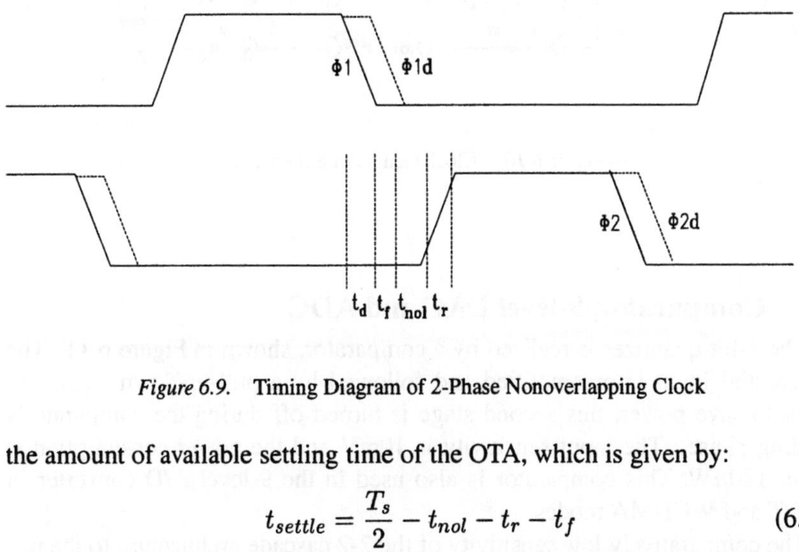

Figure 6.9. Timing Diagram of 2-Phase Nonoverlapping Clock

the amount of available settling time of the OTA, which is given by:

$$t_{settle} = \frac{T_s}{2} - t_{nol} - t_r - t_f \tag{6.15}$$

where T_s is the sampling period, t_{non} is the non-overlap time, t_r is the rise time and t_f is the fall time. In the schematic of Figure 6.10, the delay and nonoverlap times are determined by the delays in the inverter chain and the NAND gates. Because in the multi-standard application, the input clock frequency is switched among three different frequencies 13MHz, 26MHz and 39MHz, the clock generator needs to be designed based on the highest operating frequency, which is 39MHz. However in low frequency mode, the sharp rising and falling edges will generate more digital noise, which can couple to the sensitive analog part. A better way is to design three different clock generators. Each is optimized for the corresponding standard. The extra chip area associated with multiple clock generators is negligible. In our first implementation, we use the first method.

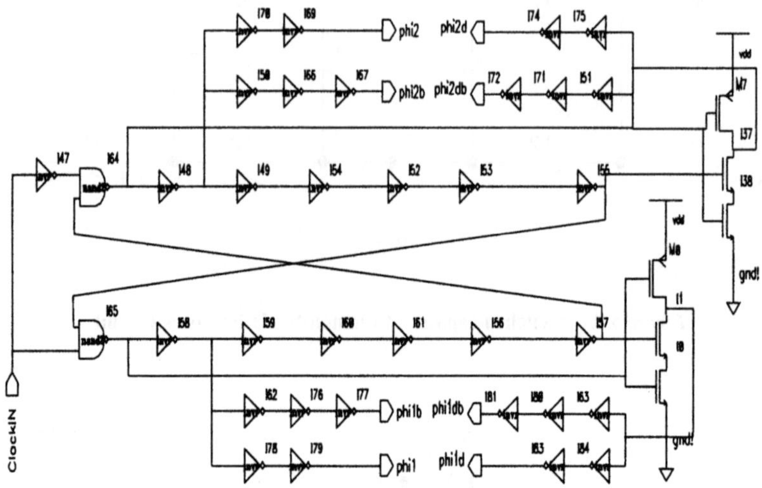

Figure 6.10. Clock Generator Schematic

5. Comparator, 9-level DAC and ADC

The 1-bit quantizer is realized by a comparator, shown in Figure 6.11. The differential input is preamplified and followed by a pull-up/down stage. In order to save power, this second stage is turned off during the comparator's holding phase. The input sensitivity is 10mV and the power consumption is about 180μW. This comparator is also used in the 9-level A/D converter in DECT and WCDMA modes.

The comparatively low sensitivity of the 2-2 cascade architecture to the imperfection of the DAC circuitry and ADC circuitry, allows use of very simple topologies for both blocks with consequent power savings. With respect to the 3-bit (9-level) ADC, a flash architecture is feasible because data has to be coded into a small number of bits at the clock rate of 39MHz. Figure 6.12 shows one unit in the flash ADC architecture. Two sampling capacitors are used to compute the difference $(V_{in+} - V_{in-}) - (V_{ref+} - V_{ref-})$. The value of the sampling capacitors is 0.5pF. At the end of phase Φ_2, the comparator is activated for evaluating the sign of the difference. The timing sequence is designed to relax the capacitive load of the fourth OTA because at the phase

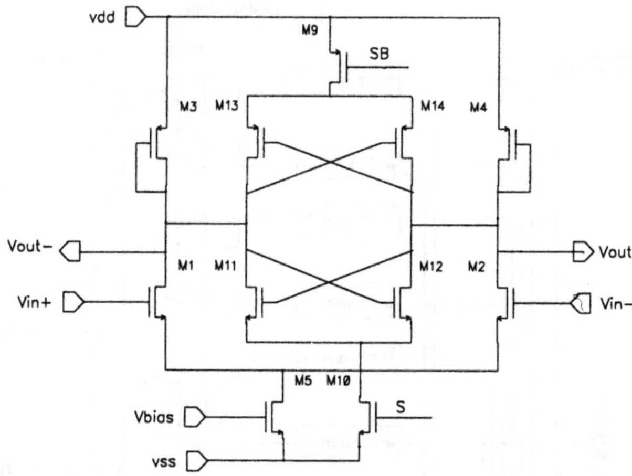

Figure 6.11. The comparator Schematic

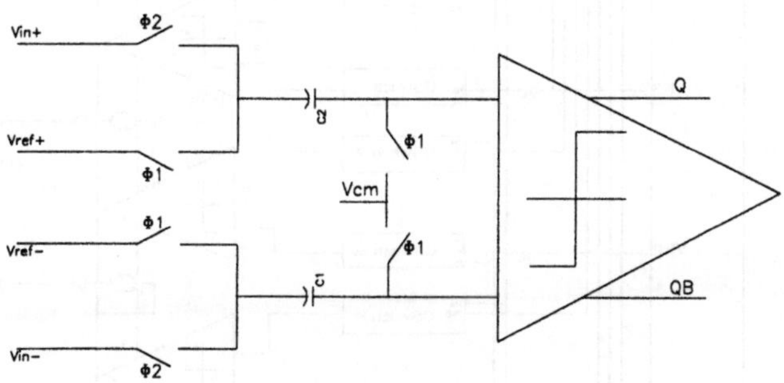

Figure 6.12. One bit Switched-capacitor Comparator

Φ_2 the load capacitor of the fourth OTA is the series capacitance of C_1 and the input capacitor of the comparator, which is much smaller than C_1.

In the flash architecture [52] [7] shown in Figure 6.13, all the comparators are identical to that used in the first- and second-stage single-bit quantizer. The reference levels are connected to the corresponding comparator inputs. In GSM mode, the mode control signal "GSMEnable" disables D8 and D0 to high and sets D1 through D7 low. The output of a separate single-bit quantizer at the second stage will set D8 and D0, thus the final output of the second-stage quantizer will be either 100000000 or 000000001. After the comparator

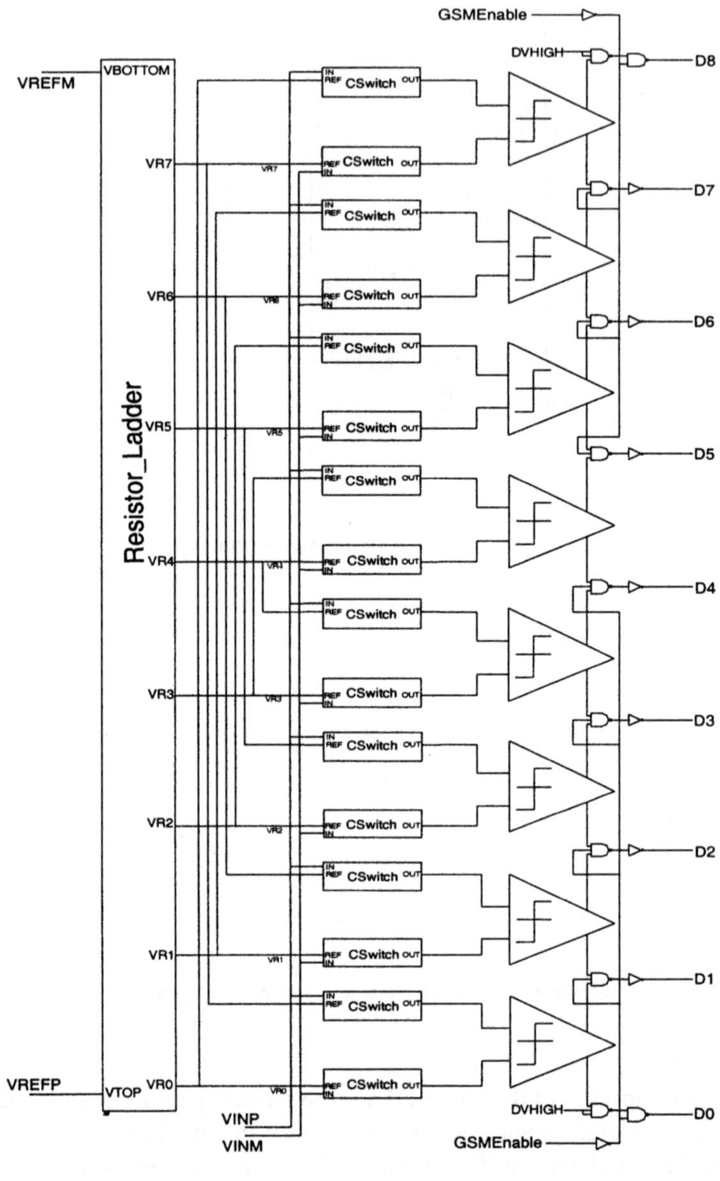

Figure 6.13. 9-level Quantizer Schematic

array, the thermal code is encoded as the 2's complement presentation, which is buffered out of the chip. The 9-level DAC is implemented by straightforward resistor-string approach. No decoder is needed since the thermal code is ready from the ADC output. To save the static power drain from the resistor string, it is shared with the reference ladder of the 9-level ADC. The 9-level DAC's nonlinearity directly contributes to the modulator's nonlinearity. According to the system simulation, performed in the previous chapter (Table 5.8), the 9-level DAC needs 8-bit linearity, or 0.4% matching. The differential and integral nonlinearity result from mismatches in the resistor comprising the ladder [60]. In the ideal case, the value of each resistor can be expressed as

$$R = \frac{\rho L}{W_t} + 2R_C \tag{6.16}$$

where ρ is the resistivity, L, W, and t are the length, width, and thickness of the resistors, respectively, and R_C represents the additional resistance due to each contact. In reality, these resistors suffer from several mismatch components: resistivity mismatch, $\delta\rho$, width, length, and thickness mismatch, δW, δL, and δt, respectively, and contact resistance mismatch, δR_C. In a typical process, δW and δL result from limited edge definition capability in lithography and etching or deposition of the resistor material, δt arises from gradients across the die, and δR_C is caused by random variations in the finite resistance at the interface of the resistor and the interconnect(usually metal). Taking the total differential of Equation 6.16, we can express the overall relative mismatch as follows,

$$\frac{\delta R}{R} = \frac{\delta\rho}{\rho} + \frac{\delta L}{L} - \frac{\delta W}{W} - \frac{\delta t}{t} \tag{6.17}$$

In a typical process, ρ and t are given, leaving L,W, and R as the only variables under the designer's control. To minimize the overall mismatch, each of these parameters must be maximized. Of course, larger dimensions lead to higher parasitic capacitance between the resistor and the substrate. Since the resistor ladder works at DC in the modulator, the parasitic capacitance is not a problem. Another important aspect of resistor value design is the Thevenin resistance seen at each reference voltage along the ladder. This resistance together with the load capacitance determines the settling time of the reference voltage. To implement the DAC, the 1-of-9 code is used to select the reference voltages generated by a resistor string. The same resistor string is used to generate the reference voltages for the ADC. This can reduce the power consumption. The resistors in the ladder are $R = 105\Omega$, which is $20\mu m \times 10\mu m$ poly2 resistor in the layout. This resistor value is low enough to ensure that the settling error of the reference voltages in the ADC input capacitors and the third and fourth integrators input capacitors are not excessive. All switches are complementary CMOS gates.

6. Digital Cancellation Circuit

The implementation of the digital cancellation circuit for Figure 6.14 also needs some consideration. Since the digital presentation of the signal is fixed point number, it is important to set the mirror between the analog data and the digital words. To simplify the digital circuits, 2's complement of the digital presentation method is used. First let us look at the one bit quantizer in the first stage. It is obvious that "1" stands for V_{ref} and "0" stands for $-V_{ref}$. However if we use only one bit word to present the output stream, there is no way to present the sum of V_{ref} and $-V_{ref}$, which should by zero. So it is necessary to encode the 1-bit quantizer output as shown in Table 6.2. So we use 2-bit digital word to present the 1-bit quantizer output. With the same reason, the 3-bit quantizer should be replaced by 9-level quantizer and the corresponding look-up table between the analog input and the encoder output is as shown in Table 6.3.

Table 6.2. 1-bit Quantizer Encoder

Analog Signal Level	Digital Words
V_{ref}	01
$-V_{ref}$	11

Table 6.3. 3-bit Quantizer Encoder

Analog Signal Level	Thermal Code Output	Encoder Output	Real Signal Level in the cancellation
V_{ref}	100000000	0100	$4V_{ref}$
$0.75V_{ref}$	010000000	0011	$3V_{ref}$
$0.5V_{ref}$	001000000	0010	$2V_{ref}$
$0.25V_{ref}$	000100000	0001	V_{ref}
0	000010000	0000	0
$-0.25V_{ref}$	000001000	1111	$-V_{ref}$
$-0.5V_{ref}$	000000100	1110	$-2V_{ref}$
$-0.75V_{ref}$	000000010	1101	$-3V_{ref}$
$-V_{ref}$	000000001	1100	$-4V_{ref}$

Based on this, the digital cancellation circuit is designed, as shown in Figure 6.14. The final two-stage combinational output signal is a 6-bit data stream. Another trick is that the multiply-by-4 for the second stage is not needed as long as, in the final adding operation, the 4-bit first stage output adds to the 6-bit second output from the LSB. The schematic of the digital cancellation circuit

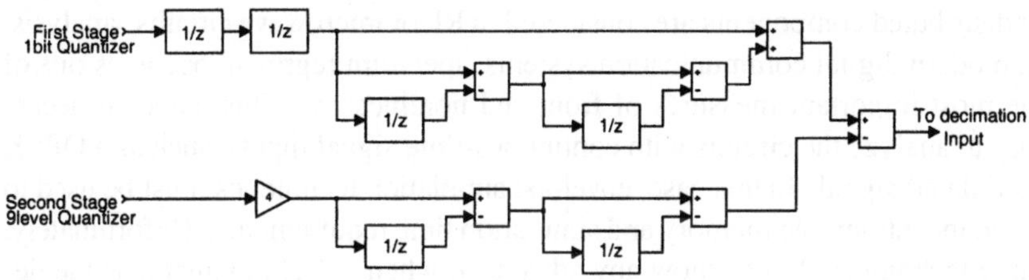

Figure 6.14. Digital Cancellation Realization Architecture

is as shown in Figure 6.15.

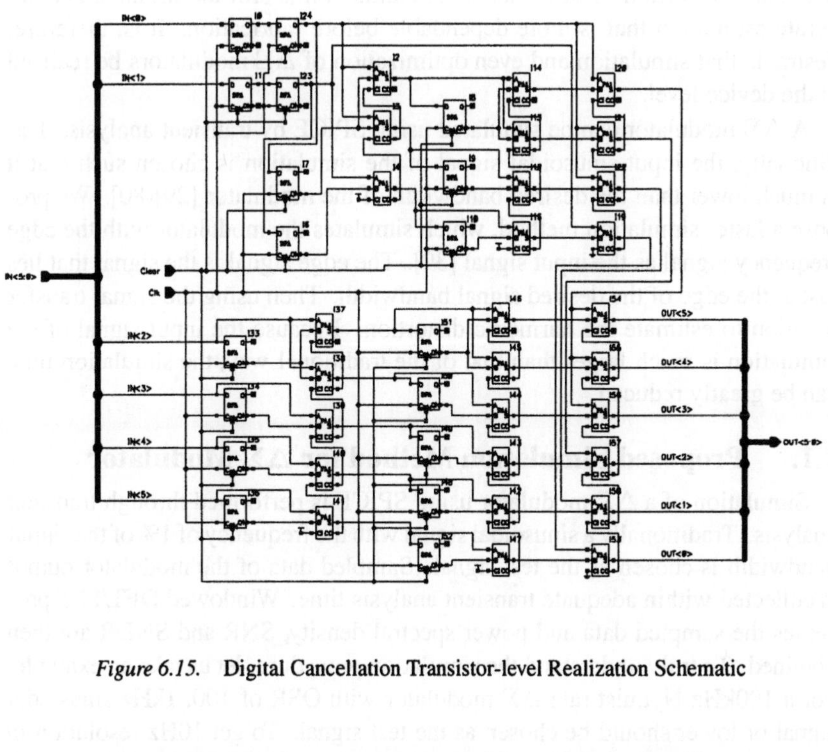

Figure 6.15. Digital Cancellation Transistor-level Realization Schematic

7. Fast Simulation of ΔΣ Modulator using SPICE

One of the most challenging aspects of designing the modulators is the simulation of performance at the device level. Besides SPICE, modern circuit CAD tools employ different techniques to deal with different signal analysis. Harmonic-balanced methods (HBM) have to be used when frequency mixing or distributed components are considered in RF or microwave circuits' analysis. In modern digital communication systems, spectrum regrowth becomes one of the most important measures of front-end nonlinearity. This makes it necessary to analyze the circuits with continuous-time signal inputs, such as a QPSK modulated signal. In this case, envelope simulation techniques must be used to avoid the intolerable memory and simulation time requirements. Unfortunately, these techniques do not show any advantage when a $\Delta\Sigma$ modulator is the design object. Thus, other simulation methods with circuit-based macromodels, time-domain macromodels, difference equations and table-lookup models are developed, such as MIDAS [87], SWITCAP2 [83], etc. These methods are very effective with mixed digital and analog sampled data systems simulation. For rapid evaluation, the use of approximate models can help reduce the number of iterations in the simulation cycle by providing good estimates for stability, SNR, and distortion. But device-level simulation is still the ultimate and accurate estimation that is more dependable before fabrication. It is, therefore, desirable that simulation and even optimization of $\Delta\Sigma$ modulators be realized at the device level.

A $\Delta\Sigma$ modulator can be simulated using SPICE by transient analysis. Traditionally, the input sinusoidal signal of the simulation is chosen such that it is much lower than the desired bandwidth of the modulator [29][80]. We propose a faster simulation method, which simulates the modulator with the edge frequency signal as the input signal [39]. The edge signal is the signal that lies just at the edge of the desired signal bandwidth. Then using the signal transfer function to estimate the harmonic distortion. Because the input signal of the simulation is much larger than that of the traditional way, the simulation time can be greatly reduced.

7.1. Proposed Simulation Method for $\Delta\Sigma$ Modulator

Simulation of a $\Delta\Sigma$ modulator using SPICE is performed through transient analysis. Traditionally a sinusoidal signal with the frequency of 1% of the signal bandwidth is chosen as the test signal. Sampled data of the modulator output is collected within adequate transient analysis time. Windowed DFT/FFT processes the sampled data and power spectral density, SNR and SNDR are then obtained. To truly understand the simulation procedure, let us take an example. For a 100kHz Nyquist rate $\Delta\Sigma$ modulator with OSR of 100, 1kHz sinusoidal signal or lower should be chosen as the test signal. To get 10Hz resolution in spectrum analysis, $10MHz/10Hz = 1Meg$ sample points should be stored. Furthermore, the convergence of SPICE requires as much as 20 times more points to be calculated. For a Pentium II/300MHz/64M RAM personal computer, one point takes approximately one second, the whole simulation will last for 1000000/3600=300hours without including the time needed for DFT/FFT. Approximately 10 days are required to acquire a single data point on the SNR-versus-input-amplitude plot! It might be faster to fabricate the chip and test it!

Since the simulation time is directly determined by the sampling rate and the test signal frequency and since the sampling rate can not be changed in the simulation because the high frequency behavior of the building blocks is hard to predict, the only way to decrease the simulation time is to increase the frequency of the test signal. If an edge frequency signal is chosen, where the edge signal means the signal lies at the edge of the signal bandwidth the simulation time be decreased dramatically. In the above example, the edge frequency of 50KHz is chosen and the simulation time is reduced by 50 times. Then the time of simulation can be tolerable.

However, if the edge signal is used as the test signal, the harmonics lie outside of the bandwidth and the SNDR cannot be estimated. So methods must be found to consider the inaccurate result of harmonic distortion.

7.2. Harmonic Distortion Analysis

A second-order switched-capacitor ΔΣ modulator is chosen to be the example to analyze the harmonic distortion behavior. The diagram of a second-order ΔΣ modulator is shown in Figure 6.16. It is straightforward to use the follow-

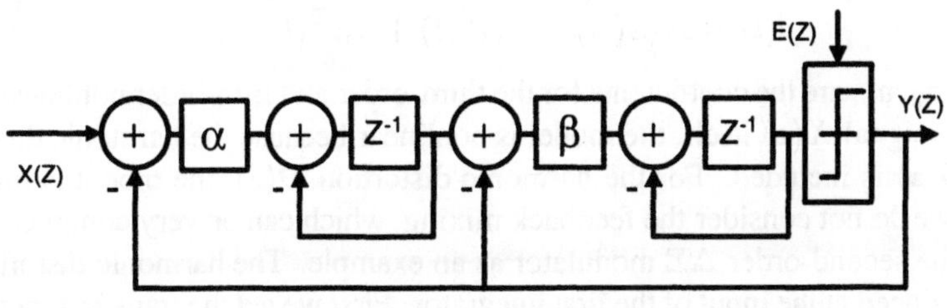

Figure 6.16. Second-order ΔΣ Modulator Linear Model

ing analysis to other higher-order ΔΣ modulators. In the following analysis, it will be assumed that the modulator is constructed with fully differential circuitry. This will result in the cancellation of all even-order distortion terms, regardless of the cause of the distortion. Signal nonlinearity is introduced by voltage-dependent capacitors, signal-dependent switch charge injection, nonlinear op-amp dc transfer function, slew rate limiting and nonlinear setting of the op-amp, nonlinear resistance of the MOS switches, and etc. [74]. It can be divided into two kinds of sources of nonlinearity. One is generated from the input of the integrator and the other is introduced at the output of the integrator. To see their effect that has on the modulator output, the distortion should be referred to the input [74]. Then the nonlinearity coming from the output of the first integrator including the nonlinear op-amp dc transfer function, slew rate limiting and nonlinear setting of the op-amp should be divided by the transfer function of the integrator. This shaping effect will cause the nonlinearity from the output of the integrator to be much less problem than the nonlinearity from the input of the integrator. So the simplified nonlinear model of the $\Delta\Sigma$ modulator shown in Figure 6.17 is proposed to analyze the harmonic distortion. Where

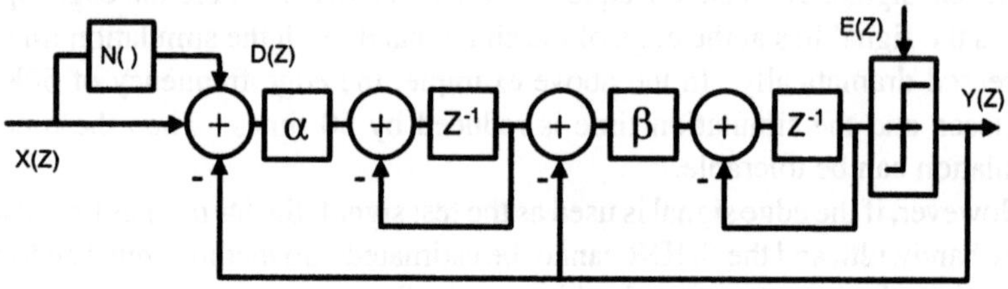

Figure 6.17. Simplified Second-order $\Delta\Sigma$ Modulator Nonlinear Model

$D(z)$ represents the harmonic distortion components referred to the input and $N()$ stands for the nonlinear function between $D(z)$ and $X(z)$. If we further assume that the nonlinearity is frequency-independent, the function of $N()$ can be expressed as:

$$d(t) = N(x(t)) = a_3 x^3(t) + a_5 x^5(t) + ... \tag{6.18}$$

where a_3, a_5, are the coefficients for the third-order and fifth-order nonlinearity. For the signal $X(z)$ itself, the model is nonlinear because the harmonic distortion $D(z)$ is included. For the harmonic distortion $D(z)$, the model is linear since we do not consider the feedback mixing, which can be very complicated. Take the second-order $\Delta\Sigma$ modulator as an example. The harmonic distortion is introduced at the input of the first integrator. First we get the transfer function from this point to the output. If the harmonic distortion component is noted as $D(z)$:

$$H_D(z) = \frac{Y(z)}{D(z)} = \frac{\alpha\beta}{(z-1)^2 + \beta(z-1) + \alpha\beta} \tag{6.19}$$

where α, β are the gains of the two integrators and $z = e^{j\omega}$. Considering two sinusoid signals with the frequency ω_1 and ω_2 respectively and the same amplitude A excite the modulator separately. The output can be expressed as:

$$Y(z) = Y_X(z) + Y_D(z) + Y_E(z) = H_X(z)X(z) + H_D(z)D(z) + H_E(z)E(z) \tag{6.20}$$

where $H_X(z), H_D(z), H_E(z)$ are the transfer functions for the signal, harmonic distortion and quantization noise. Using the model of Figure 6.17, it is obtained

$$\frac{Y_X(j\omega_1)}{Y_X(j\omega_2)} = \frac{H_X(j\omega_1)}{H_X(j\omega_2)} \tag{6.21}$$

$$\frac{Y_D(j\omega_1)}{Y_D(j\omega_2)} = \frac{H_D(j\omega_1)}{H_D(j\omega_2)} \tag{6.22}$$

$$\frac{Y_E(j\omega_1)}{Y_E(j\omega_2)} = 1 \tag{6.23}$$

Then we get the procedure to simulate the second-order $\Delta\Sigma$ modulator using SPICE at any sinusoid input signal with frequency of ω_t.

1 Choosing the edge frequency signal ω_e as the test signal and simulate the modulator by transient analysis of SPICE;

2 Using windowed FFT, find the inband noise power $|Y_E(j\omega_e)|^2$, signal power $|Y_X(j\omega_e)|^2$ and the harmonic distortion power $|Y_D(j\omega_e)|^2$;

3 Compute these power levels assume ω_t is the input:

$$|Y_E(j\omega_t)|^2 = |Y_E(j\omega_e)|^2 \tag{6.24}$$

$$|Y_X(j\omega_t)|^2 = |Y_X(j\omega_e)|^2 \frac{|H_X(j\omega_t)|^2}{|H_X(j\omega_e)|^2} \tag{6.25}$$

$$|Y_D(j\omega_t)|^2 = |Y_D(j\omega_e)|^2 \frac{|H_D(j\omega_t)|^2}{|H_D(j\omega_e)|^2} \tag{6.26}$$

4 Calculate SNR and SNDR:

$$SNR = \frac{|Y_X(j\omega_t)|^2}{|Y_E(j\omega_e)|^2} \tag{6.27}$$

$$SNDR = \frac{|Y_X(j\omega_t)|^2}{|Y_E(j\omega_e)|^2 + |Y_D(j\omega_e)|^2} \tag{6.28}$$

In the above analysis, we do not actually try to distinguish all the nonlinear sources and their respective contributions to the output. We only need to know the total effect referred to the input by one fast simulation. Ideally, quantization noise distribution and the internal noise of building blocks are independent on the input signal. For the example above, if an edge frequency signal is used as the test signal and 50 periods are required to get enough resolution, $10MHz/100KHz/50 = 5000$ sample points should be stored. The whole simulation reduced to only $5000/3600 = 1.5$hours. Here we propose a faster simulation method for $\Delta\Sigma$ modulator using SPICE. We change the time-consuming simulation problem into a much less time-consuming problem plus a little analysis work. This makes it possible that the SPICE simulation can be used to optimize the parameters in the modulator within tolerable time. Rounding and truncation errors within the circuit simulation algorithms typically set an upper limit on the measurable SNR on the order of 90dB. This implies that very high-resolution modulators cannot be correctly simulated by SPICE. Fortunately, 90dB resolution is enough for the up-date modern communication base-band A/D converter.

8. Prototype and Simulation Results

The programmable $\Delta\Sigma$ modulator is designed and implemented with a $0.35\mu m$ CMOS process. The chip area is $2mm \times 2mm$. The simulation result of the modulator for DCS1800 mode after back annotation is shown in Figure 6.18. Note that there is the product of the third harmonic, however it is well below -85dBc relative to the signal. The overall SNR is 81dB and SNDR is 79.7dB. Considering the limited Cadence computation resolution and the computation quantization noise, the modulator performance basically reaches what we expect. Figure 6.19 is the SNR versus input amplitude curve. The dynamic range is more than 90dB. The total current is 5.6mA. The power consumption is 16.8mW with 3V supply.

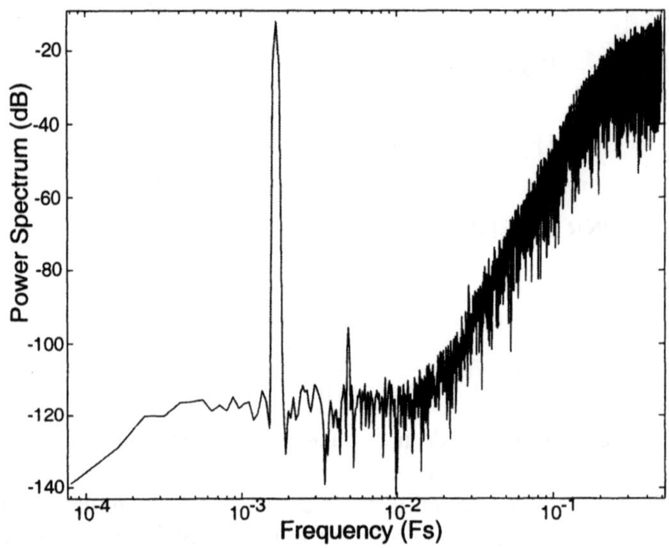

Figure 6.18. DCS1800 Mode Output Spectrum: $f_{in} = 25KHz$, $f_s = 12.8MHz$, Input Amplitude 1Vpp

9. Summary

The chapter presented the implementation of the switched-capacitor programmable $\Delta\Sigma$ modulator. The simulation of the circuits is setup in such a way that the real working environment is taken into account. The intermodulation simulation of the MOS switch shown in Figure 6.5 is an example of such simulation. A method for transistor-level simulation of $\Delta\Sigma$ modulators is also presented. This method helps to reduce the simulation time of characterizing

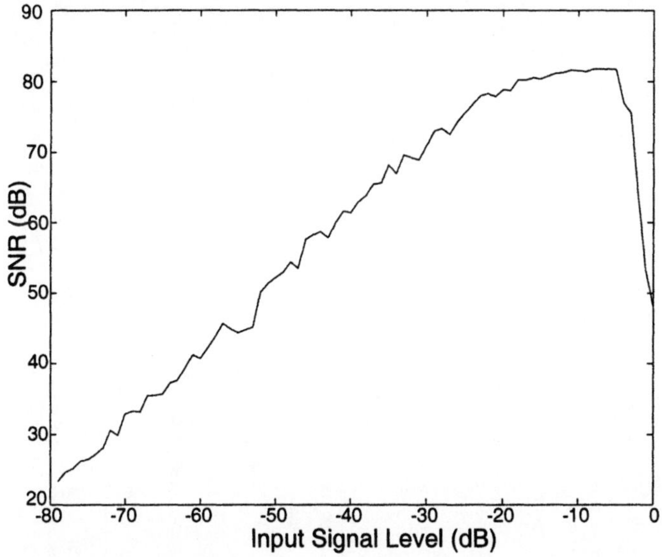

Figure 6.19. DCS1800 Mode SNR versus. Input Amplitude: $f_{in} = 25KHz, f_s = 12.8MHz$

the SNR performance of the modulator, thus allowing the possibility of circuit optimization at the top level.

Chapter 7

A/D CONVERTER TECHNIQUES FOR MULTI-STANDARD BASE-STATIONS

The capability to reconfigure transceiver systems to service both legacy services and new advanced services is an alluring goal, which will be well served by software radio architectures. Additionally, digital radios can serve wide communication bandwidth by digitally tuning to the band of interest without the need to replicate receiver circuitry on a channel-by-channel basis. A wideband digital receiver is a feasible software radio approach. In stead of an analog second IF downconversion in a superheterodyne receiver, the signal after the first IF downconversion is first converted into a digital signal by a high-speed A/D converter and the second downconversion is completed by the following programmable down converter. The programmable down converter can not be included in DSP processor because the fastest DSP processors do not yet have the bandwidth or speed necessary to process the rates at which A/D converters digitize an IF signal.

The software radio concept leads to multicarrier (multichannel) base stations for cellular applications. This creates challenges and opportunities for A/D converter designers. Traditional cellular base-station architecture uses a complete receive-and-transmit chain for each RF carrier they process. There is one antenna, but there are multiple receive chains, and each has amplifiers, filters and mixers, and usually two or three mixing stages down to baseband signal that's then digitized and goes into a DSP for the final demodualtion. To reduce space, power consumption, and most importantly, system cost, base-station designers are turning to multi-carrier radio designs that eliminate much of the redundancy of traditional narrowband designs and replace the multiple receive chains with a single high-performance radio in which each RF carrier is processed in the digital domain. By eliminating multiple analog circuits, the multicarrier approach promises to dramatically reduce base-station cost.

In a multicarrier system, the whole band is received and the digitized. All the filtering and channel selection is done digitally. Thus the A/D converter needs a lot of dynamic range. Applications such as DCS1800, PCS1900, WCDMA and WLL (wireless local loop) demand a dynamic range approaching 100dBc. Pipelined A/D converter is the most promising candidate for this application. However, the limited analog matching accuracy has been the major obstacle for high-speed high-resolution A/D converters. This chapter presents a fast and accurate calibration method that can relax the analog circuits' accuracy while putting more work in the digital domain, which can be scaled in terms of speed and power with sub-micron processing technology advancement.

1. Pipeline ADCs

A pipelined ADC is a generalized two-step ADC with improved throughput and tolerance to comparator errors. They add sample-and-hold and amplifier circuits to each stage, which allows each stage to be immediately used to process its input sample before the succeeding stages have finished. The basic architecture of a pipelined A/D converter is shown in Figure 7.1. Due to the pipelining architecture, the throughput rate is independent of the number of stages used, and the hardware cost is approximately linear with resolution. Another advantage of pipelined architectures is that the resolution requirement for the later stages is greatly relaxed compared to the first few stages, which makes it possible to reduce total power consumption. Furthermore, digital correction techniques can be used to significantly reduce the sensitivity of the architecture to analog component non-ideality. Pipeline ADCs are very suitable for high-speed and high-resolution when low power is also a major consideration. It is therefore not surprising that most of the state-of-the-art high-resolution high-speed ADCs employ pipelined architecture.

High-resolution and high-speed operation requires fast settling circuits to ensure maintenance of the resolution at high clock rates. High resolution, even at dc and low frequencies, in addition requires some form of matching enhancement to increase the matching performance of integrated circuit components such as resistors and capacitors, as well as circuit architectures which can utilize components matched to such high accuracy. Several techniques have been employed in the literature to enhance the resolution of ADCs for high performance applications, such as trimming of the unit elements [43] [10], digital domain calibration [38] [73] etc.

A new way of thinking to implement a high performance ADC is to relax the design of the analog part while using digital logic to calibrate the output to the required performance. In [8], a method was proposed to achieve this goal. The concept behind the proposed solution is to use an imprecise but fast-settling analog circuit in the ADC and to measure, during the calibration period, its entire transfer function characteristic, in general, so as to obtain knowledge

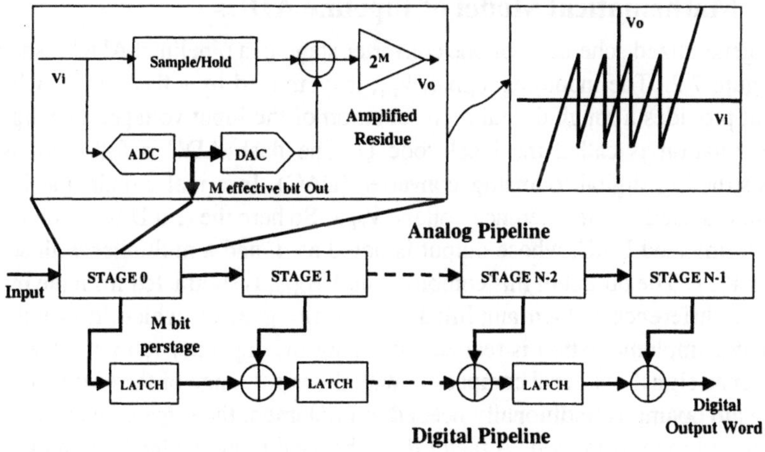

Figure 7.1. Pipelined A/D Converter Generic Architecture

of all the nonlinearities relevant at the resolution desired for the application. If the transfer function is known, and in particular if it represents a unique mapping from the input to the output, than the output can be mapped back to deduce the correct input when the ADC is in normal operation. The unique mapping property is crucial since if there is more than one input that gives the same digital output, it will not be possible to distinguish between them during normal operation. Monotonicity in the transfer characteristic, when combined with sufficient overall gain (slope) at all points, guarantees the unique mapping property, and is a convenient design goal in practice, although non-monotonic characteristics may also satisfy the unique mapping property. Generally the "monotonic+enough gain" design is more convenient.

The idea is used for the pipelined ADC implementation in [8]. The detailed analysis shows that the unique mapping can be guaranteed by deliberately adjusting the circuit parameters, such as stage gains, reference voltages while these adjustments need only 1 to 5 percent accuracy for a 14bit accuracy ADC. In particular the gain requirement on the opamps is much reduced, making possible the use of faster settling, high bandwidth and low gain opamps, such as simple differential pairs. However, the use of low gain opamps makes the circuit performance more susceptible to temperature fluctuations. Thus recalibration at periodic intervals is necessary to account for such fluctuation. It is not practical to repeatedly sweep the full input range to get the look-up table, since this takes a long time to complete and the calibration speed may be slower than the gain fluctuation speed. Thus a fast and efficient calibration method must be developed.

2. Mathematical Model of Pipeline ADCs

A generalized schematic of one converter stage in a pipelined ADC is shown in Figure 7.2. The incoming signal, V_{in}, is quantized by a flash ADC, whose output provides a rough digital representation of the input voltage. The digital representation is called the local code D. The flash ADC is followed by a reconstructing digital-to-analog converter (DAC). In a real circuit, the DAC output is scaled by the reference voltage V_{ref}. So here the real DAC is modeled as a normalized DAC, whose output is noted as d, and a multiplier with scale factor V_{ref}. The output of this combination, V_{DAC}, is subtracted from the input and the difference is then amplified. The stage gain G, which in switched-capacitor implementation is realized by sizing the capacitor ratio, restores the difference signal to a level compatible with the input range of the next stage. A high gain opamp is traditionally needed to implement the stage gain accurately in a closed loop formation. A factor θ can be used to model the limited opamp gain effect. It is a coefficient less than one. The resulting signal of the single converter stage is called the residue, V_{res}. The flash ADC's digital output D also goes to a digital error correction circuit to generate the final digital word. Mathematically, we can express the transfer function of the converter stage as:

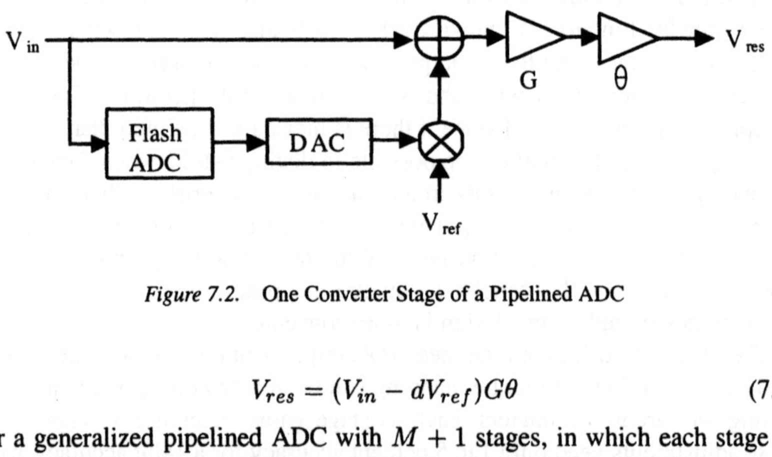

Figure 7.2. One Converter Stage of a Pipelined ADC

$$V_{res} = (V_{in} - dV_{ref})G\theta \tag{7.1}$$

For a generalized pipelined ADC with $M + 1$ stages, in which each stage of the first M stages has an effective n-bit flash sub-ADC after digital correction, the expression for the l-th stage can be specified as:

$$V_l^{res} = (V_{l-1}^{res} - d_l V_l^{ref})2^n g_l \theta_l \tag{7.2}$$

in which the reference voltage is $V_l^{ref} = x_l V^{ref}$ and $l = 1, 2, ..., M$. V_{ref} is the nominal reference voltage and x_l is a scale factor of this stage, whose value is approximately unity, representing the inaccuracy of the reference voltage. From the structure of the pipeline, it is clear that:

$$V_0^{res} = V_{in} \tag{7.3}$$

The $M-th$ pipelined stage is followed by a flash ADC, which is the $(M+1)th$ stage in our description. The output of this stage is denoted in the following as d_{flash}. For simplicity, let

$$A_l = 2^n g_l \theta_l \tag{7.4}$$

$$B_l = 2^n x_l g_l \theta_l V_{ref} \tag{7.5}$$

The expression for the input voltage can then be expanded to:

$$V_{in} = \frac{d_1 B_1}{A_1} + \frac{d_2 B_2}{A_1 A_2} + \frac{d_3 B_3}{A_1 A_2 A_3} + ... + \frac{d_{M-1} B_{M-1}}{A_1 A_2 ... A_{M-2} A_{M-1}}$$

$$+ \frac{d_M B_M}{A_1 A_2 ... A_{M-1} A_M} + \Delta \tag{7.6}$$

in which Δ stands for the quantization noise. It is less than 0.5 LSB. This requirement does not mean that the analog part of ADC must have 0.5LSB accuracy. Actually the design of the analog part of the ADC can be greatly relaxed as long as it satisfies equation 7.6 [8]. To further simplify the equation, let:

$$C_1 = \frac{B_1}{A_1} \tag{7.7}$$

$$C_2 = \frac{B_2}{A_1 A_2} \tag{7.8}$$

$$....... \tag{7.9}$$

$$C_M = \frac{B_M}{A_1 A_2 ... A_{M-1} A_M} \tag{7.10}$$

$$C_{M+1} = \frac{1}{A_1 A_2 ... A_{M-1} A_M} \tag{7.11}$$

and we get:

$$V_{in} = f(d_1, d_2, ..., d_M, d_{M+1})$$

$$= \sum_{i=1,M+1} d_l C_l = d_1 C_1 + d_2 C_2 + ... + d_M C_M + d_{flash} C_{M+1} \tag{7.12}$$

Note that we eliminate the quantization noise in equation 7.12. Strictly to say, the equation is not accurate, but it is accurate enough to hold the required resolution since the quantization noise is less than 0.5LSB. We call this equation as the pipelined ADC's ideal inverse transfer function because it describes how to calculate the input of the ADC from its digital output. In equation 7.12 the original digital outputs of the flash sub-ADCs are used, not the digital word after error correction. Actually, equation 7.12 itself already includes the traditional error correction. In the generalized equation, x_l, g_l, θ_l can be different for different stages. They presents the reference error, gain error and limited opamp gain effects. These factors are assumed to be linear, which means they are signal invariant. Although input voltage V_{in} is linearly related to C_1~C_{M+1} the relationship between V_{in} and x_l, g_l, θ_l is very complex and nonlinear. Traditionally, the digital outputs of each stage, d_i, are combined to form the digital conversion code. Unfortunately, if x_l, g_l, θ_l are not ideal (equal to 1), the combined output codes will not be accurate, generating large INL and DNL. However, since we have the inverse transfer function model shown as equation 7.12, if we can obtain C_1 through C_{M+1}, the input signal can be accurately regenerated based on the original digital outputs from each converter stage.

Using equation 7.12 to accurately recover the original signal, there is one other necessary condition. In equation 7.12, given a set of output code d_1 through d_{M+1}, only one input value must be mapped. In other words, if the ADC converts two input signals, whose difference is greater than 1 LSB, into the same code, we cannot use equation 7.12 any more. So some design method should be used so that it can be guaranteed that the mapping from the input signal to the output code never be multiple-to-one. There may be a lot of ways to avoid the multiple-to-one mapping ADC characteristic. One such method for the case of a switched-capacitor pipelined ADC is proposed in [8].

3. A Fast Calibration Algorithm for Pipeline ADCs

Based on the generalized model of a pipelined ADC, a fast calibration algorithm can be developed. To describe the calibration algorithm, we take a 14-bit 6-stage pipelined ADC as an example. For the first five stages, each stage has a 2.5-bit flash sub-ADC. The last stage is a 4-bit flash converter. From equation 7.12, it is easy to find that if we have $M+1$ points measured, which means we have V_{in} and d_1 through d_{M+1} for these points, a square matrix $(M+1) \times (M+1)$ equation can be obtained.

$$(V_{in})_j = \sum_{i=1,M+1} C_i(d_i)_j j = 1, 2, ..., M, M+1 \qquad (7.13)$$

The unknown C_1 through C_{M+1} can be obtained by solving this system of linear equations. In practice, to make sure that the measured points includes all the information from each converter stage and to make the algorithm more robust to the measured points, a better way to find the accurate values of C_1 through C_{M+1} is to measure additional points and thus obtain more equations and then solve an over-determined least squares problem. So in equation 7.13, it is not necessary to limit j from 1 to $M+1$. For this example, M is 5 and a total of 12 points chosen at random from the entire conversion range are enough for the 14-bit accuracy based on the simulation. A diagram of the implementation

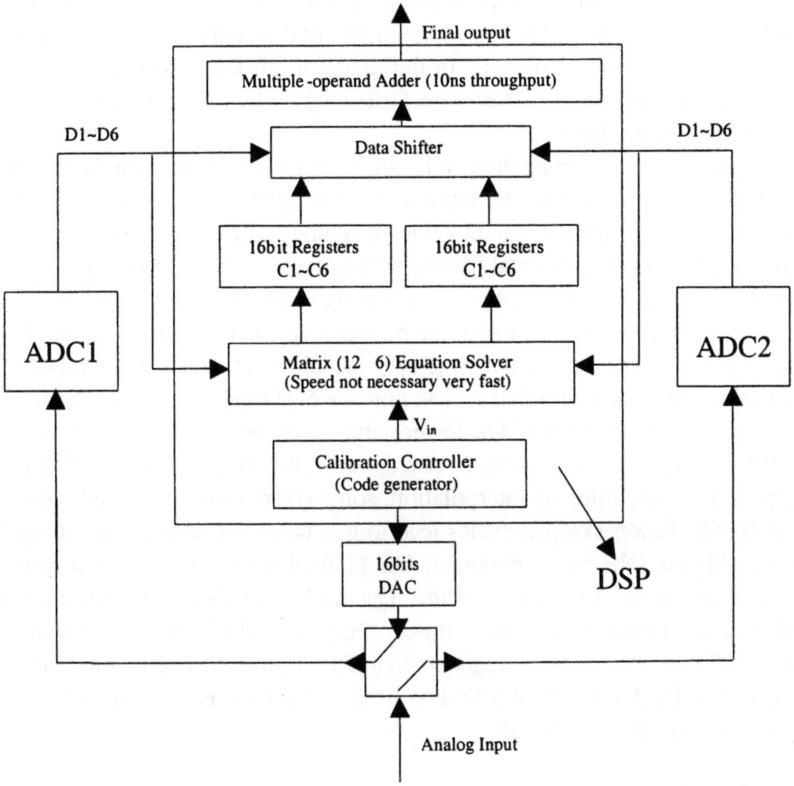

Figure 7.3. Background Fast Calibration Implementation for 14-bit 100Mbps 6-stage Pipelined ADC

using a ping-pong background calibration method for a 14-bit 100Mbps 6-stage pipelined ADC is shown in Figure 7.3. In this implementation, two ADCs are needed. One runs in calibration mode while the other runs in normal signal conversion mode. The ADCs are periodically switched in ping-pong fashion as required to maintain the calibration updated. This gives background operation at the expense of doubling the ADC hardware. Because the digital outputs (in the example d_1~d_6) are only fixed values, a fully functional multiplier is not needed. Only data shifters and an adder are needed to complete the function of multiplication. Of course the throughput speed of the data shifters together with the adder should be the same as the ADC's throughput speed. Registers are required to store the coefficients, C_1~C_6 in the example, which are the results of the equation solver. When the ADC is in calibration mode, a calibration controller (code generator) generates some input codes to a 16-bit calibration

DAC and the output of the DAC is fed to the ADC. Both the calibration codes and the output codes d_1˜d_6 are input to the matrix equation solver and the coefficients are then calculated. In the diagram, all the blocks in the dotted square can be realized in a separate DSP chip. On-chip realization is also possible and straightforward.

Compared with sweeping the whole range with a 16bit DAC to update a look-up table and check the look-up table to get the calibrated output, this method takes advantage of the regular pipeline structure and is much faster and easy to implement. Instead of establishing a large look-up table, which consumes considerable memory, the calibrated output is calculated with fast registers and an adder. Another possibility for background operation is a skip-and-fill type mode of operation similar to [37] in which input cycles are skipped for calibration and later interpolated. The number of skipped cycles is determined by the calibration algorithm. The fewer points to be measured in the calibration procedure, the fewer cycles need to be skipped in the ADC operation. When the skipped cycles are filled by interpolation, some error or noise is added onto the input signal. Fewer skipped cycles lead to less noise added to the input signal and thus higher SNR can be obtained. Our method is advantageous in requiring only a small number of skipped cycles. The implementation of the skip-and-fill background calibration is even simpler. Only one ADC channel is required. Instead of putting two sets of registers in a DSP to save the coefficients for two ADC channels, the power of a DSP is used to realize the skip-and-fill timing and the interpolation algorithm.

4. DAC Nonlinearity

One of the major nonideal effects of a pipelined ADC is from the nonlinearity of the multi-bit DAC. In the generalized model of pipelined ADCs developed in previous section, the DAC nonlinearity is not included. Without considering the DAC nonlinearity in the calibration algorithm, the SNDR performance of the ADCs will be dominated by the DAC nonlinearity. For example, 0.2% mismatch in the DAC components results in 72dB SNDR for the 14-bit pipelined ADC, which is far less than the required resolution. In a practical implementation, the multi-bit DAC is realized by capacitor arrays. Figure 7.4 shows the schematic of such a 2.5-bit DAC for the 14-bit pipelined ADC. The transfer function of the $l-th$ converter stage thus can be expressed as:

$$V_l^{res} = \frac{C_{l1}+C_{l2}+C_{l3}+C_{l4}}{C_{l4}} V_{l-1}^{res} - (d_{l1}\frac{C_{l1}}{C_{l4}} + d_{l2}\frac{C_{l1}}{C_{l4}} + d_{l3}\frac{C_{l3}}{C_{l4}})V_l^{ref} \tag{7.14}$$

d_{l1}, d_{l2} and d_{l3} are the output of the $l-th$ flash ADC and $V_l^{ref} = V_{ref+} = -V_{ref-}$. In equation 7.14, d_{l1}, d_{l2} and d_{l3} can be 1, -1, 0 and the corresponding capacitor is connected to V_{ref+}, V_{ref-} or ground respectively. Ideally all the capacitors C_{l1}, C_{l2}, C_{l3} and C_{l4} are equal. The DAC nonlinearity comes from

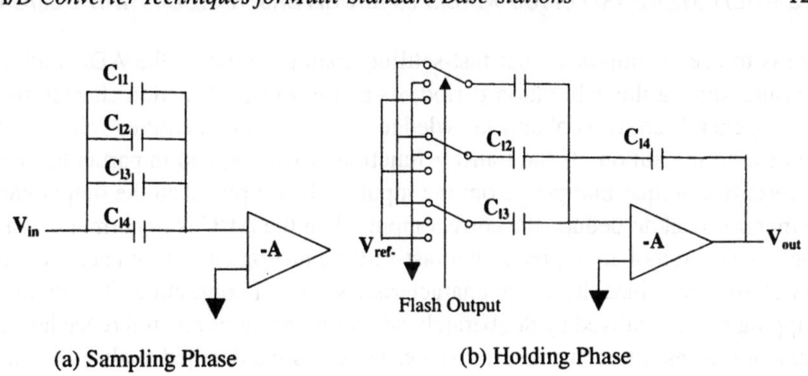

Figure 7.4. 2.5-bit Capacitor Array DAC

the mismatches among these capacitors. The mismatches have two nonideal effects for the whole ADC. One is the gain error and the other is the non-uniform reference level. In equation 7.12, the gain error can be expressed and thus be calibrated. To calibrate the DAC nonlinearity, we need to add more coefficients in the modeling. Adding equation 7.14 into equation 7.12, it is easy to get:

$$
\begin{aligned}
V_{in} &= f(d_{l1}, d_{l2}, d_{l3}, ..., d_{M1}, d_{M2}, d_{M3}, d_{M+1}) \\
&= \sum_{i=1,M} (\sum_{j=1,3} d_{ij} r_{ij}) C_i + d_{M+1} C_{M+1} \\
&= \sum_{i=1,M} \sum_{j=1,3} d_{ij} F_{ij} + d_{M+1} C_{M+1} \qquad (7.15)
\end{aligned}
$$

where $r_{ij} = \frac{C_{ij}}{C_{i4}}$, $F_{ij} = r_{ij} C_i$ and d_{ij} is the output of the i-th flash ADC. Now there are $3M+1$ coefficients in the inverse transfer function to be solved, if a 2.5-bit DAC is used. For the 14-bit pipelined ADC example, 16 coefficients are to be calculated by solving a matrix equation. So the calibration speed can still be very fast. When the gains of the converter stages fluctuate with time, only the C_is are affected. To first order, it can be assumed that component mismatches never change with time. Therefore, though there are more coefficients in equation 7.15 than those in equation 7.12, the calibration complexity is not increased except for a larger matrix equation to be solved initially. As for the 14-bit pipelined ADC example, only 6 coefficients still need to be updated thereafter.

5. Calibrated Pipeline ADC Architecture

A calibrated pipelined ADC architecture using a modified transfer function approach is proposed by [8]. The concept behind the proposed ADC architecture is to use an imprecise but fast-settling analog circuit in the ADC and to measure, during the calibration period, its entire transfer function characteristic, in general, so as to obtain knowledge of all the nonlinearities relevant at the desired resolution. If the transfer function is known, and in particular if it represents a unique mapping from the input to the output, than the output can be mapped back to deduce the correct input when the ADC is in normal operation. The design of an imprecise but fast-settling analog circuit is then easy and straightforward since these two characteristics are not conflicting. The unique mapping can be realized by deliberately adjusting the gains and reference levels (capacitor sizes) of each converter stage. [8] gives the detailed analysis of this technique. In designing an ADC using the proposed technique, some additional practical considerations arise as also described in [8]. One consideration concerns stabilizing the temperature dependence of the circuit. The use of low gain opamps as proposed makes the circuit performance more susceptible to temperature fluctuations. Thus recalibration at periodic intervals is necessary to account for such fluctuation. The algorithm developed so far is suitable for this application, without requiring the more complex analog circuitry proposed in [8] for correcting such fluctuations.

Table 7.1. Behavioral Modeling of Errors in Pipelined ADCs

Modeled Errors	Circuit Nonideality	Modeling Parameters	Drifting Speed
Residue amplifier gain	Component mismatch	g_i	~Zero
ADC threshold	Comparator offsets	Not needed	Very fast
Reference voltage error	Voltage source error	x_i	~Zero
Nonlinear DAC	Component mismatch	r_{ij}	~Zero
Opamp gain error	Opamp limited gain	θ_i	Slow

The calibration method is simulated with the 14-bit 6-stage pipelined ADC using a behavioral model. The error sources of the ADC are modeled in the simulation as shown in Table 5. The table shows the circuit nonidealities, which are the physical sources of these errors. All these errors are mapped into the corresponding parameters in the generalized ADC model in equation 7.12 and 7.15. Though comparator offsets can change very fast, the problem is solved by the digital error correction technique, which is inherently included in the generalized model. Component problems, such as residue amplifier gain error, reference voltage error, DAC nonlinearity, do not change with time and temperature; and thus only one calibration run is needed to correct these errors. The most important error source in the calibration method is the opamp gain error. In order to relax the design trade-off between gain and speed, the opamps are designed with fast speed and low gain. Thus the gain drifts slowly with the temperature. It is required to update the drift in the calibration from time to time. To verify our calibration method, the simulation procedure is as follows:

1 Set the nonideal parameters x_l, g_l, θ_l, and capacitor mismatches for all the stages.

2 Generate some codes for the calibration DAC and measure the raw output code from the ADC.

3 Establish and solve matrix equations by DSP and get the 16 coefficients in equation 7.15

4 Input a sine signal to ADC and calibrate the output using equation 7.15.

5 Change the opamp gains by setting new θ_l values. Repeat 2 and 3

Once we have the 16 coefficients, we only need to update 6 coefficients in the following calibration, which means we still only need to measure tens of points to solve 6 coefficients instead of 16. This can be easily recognized in equation 7.15 since θ_l only affects C_i.

Table 7.2. A/D converter Calibration Simulation: Parameter Set 1

Stage	x	θ	G	c_{i1}	C_{i2}	C_{i3}
1	0.99	0.99	1.05	1.001	0.999	1.001
2	0.98	0.98	1.05	1.000	0.999	1.001
3	0.97	0.97	1.05	0.999	0.999	1.001
4	0.96	0.96	1.06	1.001	0.999	1.000
5	0.99	0.95	1.05	1.001	1.000	1.001

Table 7.3. A/D converter Calibration Simulation: Parameter Set 2

Stage	x	θ	G	c_{i1}	C_{i2}	C_{i3}
1	0.99	0.98	1.05	1.001	0.999	1.001
2	0.98	0.99	1.06	1.000	0.999	1.001
3	0.97	0.98	1.05	0.999	0.999	1.001
4	0.96	0.97	1.06	1.001	0.999	1.000
5	0.99	0.98	1.05	1.001	1.000	1.001

Using the method proposed in [8], the parameters are set as shown in Table 7.2 and Table 7.3. Assume the nominal gains of opamps are 40dB and the gain drift with the temperature is 35dB˜45dB, which leads to θ in the range of 0.9825˜0.9944. In the simulation we set θ in the range of 0.96˜1.

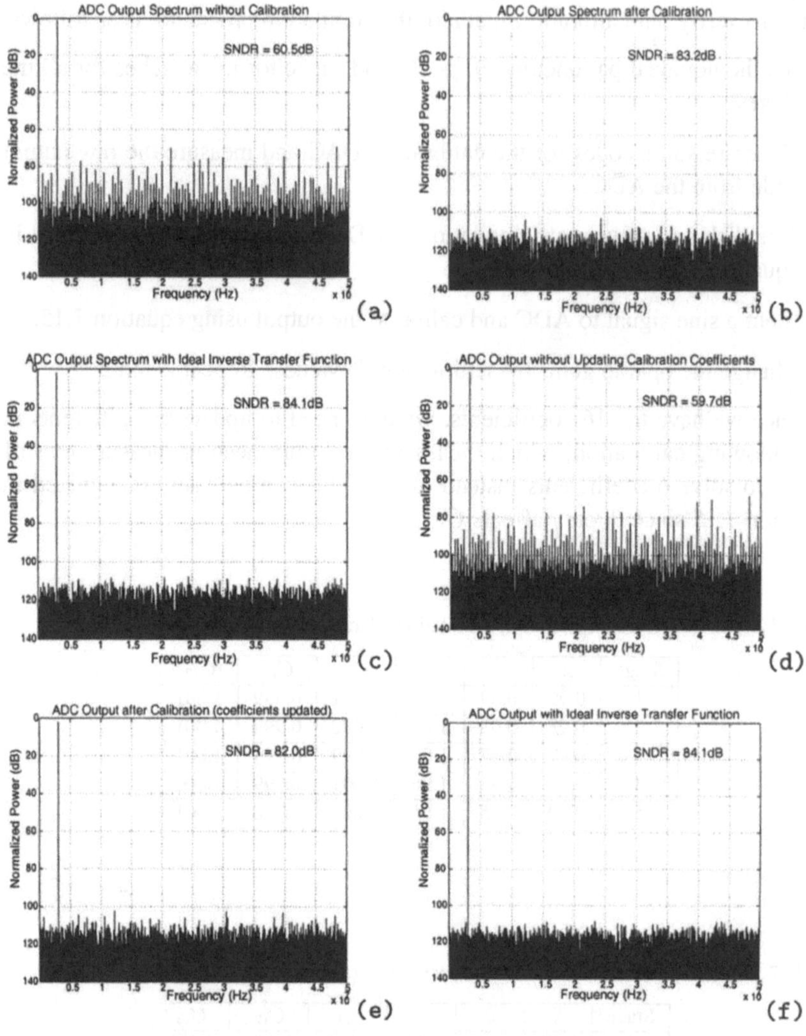

Figure 7.5. A/D Converter Calibration Simulation Results

Figure 7.5 shows the simulation results of our calibration method. The input signal frequency is 6MHz and the sampling frequency is 100MHz. The capacitor mismatch is taken to be 0.2%, which can be easily realized in standard CMOS technology. The nominal opamp gain is 40dB, which is low enough to design the high-speed characteristic without difficulty. In Figure 7.5, (a) is the original output spectrum of the ADC with parameter set 1. The SNDR is 60.5dB, which is far away from the 14-bit resolution required. 30 random points are measured and using equation 7.15, 16 coefficients are calculated. Figure 7.5 (b) shows the output spectrum after the calibration using the 16 coefficients solved. The SNDR is 83.2dB. To show the performance of the calibration method, Figure 7.5 (c) presents the calibration result using the ideal inverse transfer function of the ADC. The SNDR is 84.1dB, which is the best theoretical performance we can get from the ADC. Figure 7.5 (d) shows the output spectrum of the ADC with parameter set 2 after calibration while in the calibration the previously calculated 16 coefficients are used. The difference between parameter set 1 and set 2 is only in the opamp gains θ of all the stages. In Figure 7.5 (d), SNDR is 59.7dB, showing that updating of the coefficients is necessary. Figure 7.5 (e) is the calibrated output spectrum with the updated coefficients. Note that we use only 15 measured points to update 6 coefficients, which are the C_is in equation 7.15. The SNDR is now 82dB. Figure 7.5 (f) also shows the calibrated output using the ideal inverse transfer function.

6. Summary

In this chapter, a new calibration method is proposed to calibrate pipelined ADC architectures for high-resolution and high-speed applications. By measuring only tens of points, the whole ADC's inverse transfer function can be obtained and thus the output calibrated to the resolution required. The proposed ADC architecture in [8] together with the fast calibration method makes it possible that the original analog ADC be easily designed. The building blocks' characteristic can fluctuate with time and temperature, but the method provides for easy updating of the calibration to account for this. The main advantage of the method is that in the analog ADC design the reference voltage, the capacitor matching and the opamp gain need not be accurate. The calibration will deal with these nonidealities and correct them to the required resolution. Though in this work, we use a 14-bit 6-stage pipelined ADC as an example, the method itself is suitable for all pipelined ADCs.

Chapter 8

CONCLUSION

The proliferation of 2nd generation cellular standards and the imminent arrival of the 3rd generation cellular standard, together with other emerging wireless standards like Bluetooth and GPS, have resulted in the urgent need to develop terminals that can support multi-standard operation, which we call all-in-one phones. To ensure widespread acceptance and usage, it is important for terminals to be capable of multi-standard operation, yet meet the commercial constraints of cost, size and power consumption. This book is based on this background in mind.

Multi-band and multi-standard receiver architectures have been discussed in some papers. They mainly focus on the direct conversion receiver and wide-band double-IF receiver. However, this book is the first to provide a complete analysis of the receiver chain from the antenna to the A/D converter. Smart architecture-level management enables a compact multi-standard receiver and minimizes building blocks and thus the power consumption. Though the work targets to integrate WCDMA/DCS1800/DECT standards, it is straightforward to design the architecture for other standards. The reason to choose these three standards is that they stand for 3G, 2G and cordless phone system respectively. The receiver chain is analyzed and the specifications of every building block are given. The important thing is that these specifications are set such that it is feasible to be implemented with the current CMOS processing technology. Thus a single-chip CMOS receiver, from RF to baseband, can be fully accomplished.

The difference between multi-mode and multi-standard receiver is in the received baseband signal after downconversion. In multi-mode receivers, the downconverted baseband signals are all the same though they may be upconverted to different RF carrier frequencies in transmitters. In multi-standard receivers, not only the RF carriers are different but also the baseband signals are different. The baseband signals have different bandwidths for different standards, from 30KHz to several MHz. This brings more challenges in the receiver design. In our work, the multi-standard baseband signals are successfully processed by adopting a mixed-signal strategy that fully exploits $\Delta\Sigma$ modulation technique on channel filtering or A/D conversion. A programmable $\Delta\Sigma$ modulator A/D converter has the ability to convert different signals at different resolutions. It lends itself naturally to multi-standard wireless design. But to completely convert the WCDMA 2.5MHz signal without filtering out large blockers, would make the A/D converter power hungry [22] and it is therefore more reasonable to design such A/D converter for a base-station receiver rather than a mobile handset. A 6-th order channel filter is employed to relax the A/D converter specification for WCDMA so that only one $\Delta\Sigma$ modulator is used for the whole multi-standard receiver. Note that we set this channel filter a fixed filter, though it can be designed to be a programmable filter to relax the A/D converter requirement [5]. We do this because it is difficult to control the cut-off frequency of an analog filter, especially for narrow-band signal standard. A digital filter is much more accurate and easy to control. Another issue in analog filter design is that for some standards, like PHS, DAMPs, the signal bandwidth is so small that realizing the required capacitance and resistance in the filter will not be practical because of the huge area they occupy. It is better to put more signal processing in the digital domain rather than in the analog domain because both power and chip area will be drastically decreased as CMOS technology scales down to deep sub-micron feature sizes.

A programmable $\Delta\Sigma$ modulator A/D converter is proposed, designed and implemented. It behaves as a 14bit, 11bit and 7bit A/D converter for DCS1800, DECT and WCDMA respectively. This is achieved by innovation at the architecture level. The $\Delta\Sigma$ modulator is a 2-2 MASH with switchable 1bit/3bit last stage quantizer. The switched-capacitor implementation is realized based on thorough analysis and simulation.

In a wireless communication system, base-stations also need software radio operation to save power and adapt to the evolution of new wireless generations. High-speed high-resolution A/D converters represent the bottleneck in the design of such basestations. The main reason stems from the practical imperfection in analog circuits. Because of the matching limitation in real circuits, to get higher than 14bit resolution at 100MHz speed becomes extremely difficult. Calibration must be used to compensate for the analog inaccuracy. In our research, a fast and accurate algorithm is developed to relax the requirements of analog accuracy. A low-gain high-speed amplifier can be used instead of a high-gain, high-speed amplifiers. Less matching accuracy is required in components that determine the resolution of the converter. Now a high-speed high-resolution A/D converter is then implemented via a high-speed low-resolution A/D converter with a powerful digital signal processing ASIC. This can further extend the A/Ds' performance and hold the advantage from more digital circuitry for future processing technology.

In summary, this book focuses on multi-standard and software radio receiver design at both the system and circuit levels wherein we have:

- Developed an integrated radio front-end Zero-IF architecture for multi-standard operation of WCDMA, DCS1800 and DECT. The architecture can be easily extended to more standards, like Bluetooth, HomeRF, PDC, IS95, cdma200 etc.

- Performed system analysis of the radio subsystem and specification distribution of building blocks.

- Explored RF CMOS front-end circuit design techniques. A novel GHz LNA design technique including an internal matching inductor is presented which can benefit both NF and gain of LNAs. A new GHz prescaler used in PLL frequency synthesizer is presented and measurement results verify the validity of the architecture and design.

- Proposed, analyzed and designed a programmable 2-2 cascade $\Delta\Sigma$ modulator with multi-bit quantizer, which can meet the baseband processing requirement for WCDMA, DCS1800, and DECT multi-standard receiver. New Slew rate modeling method is discussed and fast device level simulation method for $\Delta\Sigma$ modulators is presented.

- Developed a fast calibration algorithm for high-speed high-resolution pipeline A/D converters. The method deals with the hard-to-realize analog accuracy problem by incorporating powerful digital signal processing, making it suitable for future base-station software radio receivers.

Appendix A
Acronyms

ACS : adjacent channel selectivity

ADC : analog-to-digital converter

AGC : automatic gain control

$AMPS$: advanced mobile phone system

BER : bit error rate

$CDMA$: code-division multiple access

$CT-2$: cordless telephony 2

DAC : digital-to-analog converter

DBS : digital broadcasting system

$DCS1800$: digital cellular system

DCR : direct conversion receiver

$DECT$ digital enhanced cordless telecommunications

DFF : D-flip-flop

DFT : digital fourier transform

DSP : digital signal processing

ECL : emitter-coupled logic

$EGSM$: enhanced GSM

$EDGE$: enhanced data GSM evolution

FDD : frequency division duplex

$FDMA$: frequency division multiple access

FFT : fast fourier transform

$FPGA$: field programmable gate array

$GMSK$: gaussian minimum shift keying

$GFSK$: gaussian frequency shift keying

$GPRS$: general packet radio service

GPS : global positioning system

IF : intermediate frequency

$IIP3$: input third-order intercept point

$IMT - 2000$: international mobile telecommunication 2000

ISM : industrial, science and medicine

ITU : international telecommunication union

IR : image rejection

$QPSK$: quadra-phase shift keying

GSM : global system for mobile communications

LAN : local area network

LNA : low noise amplifier

LPF : low-pass filter

LO : local oscillator

MAN : metropolitan area network

$MASH$: multi-stage noise shaping

$MEMS$: micro electro mechanical systems

NF : noise figure

OSR : oversampling ratio

OTA : operational transconductance amplifier

PA : power amplifier

PAN: personal area network

$PABX$: private automatic branch exchange

$PACS$: persoanl access communication system

PHS : personal handy-phone system

PCN : personal communications network

PLL : phase-locked loop

SAW : surface acoustic wave

$SFDR$: spurious-free dynamic range

SIR : signal-to-interferer ratio

SNR : signal-to-noise ration

SR : slew rate

SWR : software radio

$TDMA$: time division multiple access

TDD : time division duplex

THD : total harmonic distortion

$UMTS$: universal mobile telecommunications service

VGA : variable gain amplifier

VCO : voltage controlled oscillator

WAN : wide area network

$WCDMA$: wideband code-division multiple-access

$WLAN$: wireless local area network

WLL : wireless local loop

References

[1] 3GPP. "3rd Generation Partnership Project (3GPP) Technical Specification Group (TSG) RAN WG4 UE Radio transmission and Reception (TDD)". *http://www.3gpp.org*, 2000.

[2] 3GPP. "3rd Generation Partnership Project (3GPP) Technical Specification Group (TSG) RAN WG4 UE Radio transmission and Reception (FDD)". *http://www.3gpp.org*, 2000.

[3] Adiseno, M. Ismail, and H. Olsson. "A Wideband Active-Feedback Low-Noise Converter for Multiband High-Linearity Low-IF Wireless Receiver". *Proceedings of IEEE BCTM*, pages 131–134, 2001.

[4] H. A. Alzaher, H. O. Elwan, and Mohammed Ismail. "A CMOS Highly Linear Channel-Select Filter for 3G Multistandard Integrated Wireless Receivers". *IEEE Journal of Solid-State Circuits*, 37(1):27–37, January 2002.

[5] H.A. Alzaher, H.O. Elwan, and Mohammed Ismail. "CMOS Digitally Programmable Filter for Multi-standard wireless receivers". *Electronics Letters*, 36(2):133–135, January 2000.

[6] Anadigics. AWS5509S17, Tri-Band Phemt SP4T Transmit/Receive Antenna .

[7] B. Brandt and B. A. Wooley. "A 50MHz Multibit sigma-delta modulator for 12b 2Mhz A/D conversion". *IEEE Journal Solid-State Circuits*, 26(12):1746–1756, 12 1991.

[8] Alex R. Bugeja. "A Calibrated Pipelined ADC Architecture Using a Modified Transfer Function Aroach". *To be published.*

[9] Byungsoo Chang, Joonbae Park, and Wonchan Kim. "A 1.2GHz CMOS Dual-Modulus Prescaler Using New Dynamic D-Type Flip-Flops". *IEEE Journal of Solid-State Circuits*, 31(5):749–752, May 1996.

[10] M.-J. Choe, B.-S. Song, and K. Bacrania. "A 13b 40Msample/s CMOS Pipelined Folding ADC with Background Offset Trimming". *ISSCC Digest of Technical Papers*, pages 34–35, February 2000.

[11] Jan Craninckx and Michiel S. J. Steyaert. "A 1.75GHz/3V Dual-Modulus Divide-by-128/129 Prescaler in 0.7μm CMOS". *IEEE Journal of Solid-State Circuits*, 31(7):890–897, 1996.

[12] J. Crols and M. Steyert. "Low-IF Topologies for High-performance Analog Front Ends of Fully Integrated Receivers". *IEEE Transactions on Circuits and Systems II: Analog and Digital Signal Processing*, 45(3):269–282, March 1998.

[13] Erik Dahlman, PerBeming, and et al. "WCDMA-The Radio Interface for Future Mobile Multimedia Communications,". *IEEE Transactions on Vehicular Technology*, 47(4):1105–1118, 1998.

[14] Bernhard E.Boser and Bruce A.Wooley. "The Design of Sigma-Delta Modulation Analog-to-Digital Converters". *IEEE Journal Solid-State Circuits*, 23(12):1298–1308, 12 1988.

[15] Hassan Elwan, Hussain Alzaher, and Mohammed Ismail. "A New Generation of Global Wireless Compatibility". *IEEE Circuit and Devices Magazine*, pages 7–19, January 2001.

[16] Peter Weger et al. "Completely Integrated 1.5GHz Direct Conversion Transceiver". *1994 Symposium on VLSI Circuits Digest of Technical Papers*, pages 135–136, 1994.

[17] Steven R. Norsworthy et al. "A 14-bit 80-kHz Sigma-Delta A/D Converter: Modeling, Design, and Performance Evaluation". *IEEE Journal of Solid-state Circuits*, 24(2):256–266, April 1989.

[18] ETSI. "Digital cellular telecommunications system; radio transmission and reception (GSM 5.05)". *European Telecommunications Standards Institute, Copyright 1996.*

[19] ETSI. "Radio Equipment and Systems (RES); Digital European Cordless Telecommunications (DECT); common Interface (CI); Part2: Physical Layer (PHL)". *European Telecommunications Standards Institute, Copyright 1996.*

[20] Kamilo Feher and Hussein Mehdi. "Modulation/Microwave Integrated Digital Wireless Developments". *IEEE Transactions on Microwave Theory and Techniques*, 43(7):1715–1732, July 1995.

[21] A. Fernandez-Duran and et al. "Zero-IF Receiver Architecture for Multistandard Compatible Radio Systems, GIRAFE project". *Vehicular Technology Conference, 1996. Mobile Technology for the Human Race., IEEE 46th*, 2(2):1052–1056, 1996.

[22] Ichiro Fujimori and et al. "A 90dB SNR 2.5MHz Output Rate ADC using Cascaded Multibit Sigma-Delta Modulation at 8x Oversampling Ratio". *2000 IEEE International Solid-State Circuits Conference Digest of Technical Papers*, pages 338–339, Feb 2000.

[23] Thomas Gee. "Suppressing Errors in W-CDMA Mobile Devices ". *Communicaiton System Design*, 7(3):33–38, 2001.

[24] Paul R. Gray and Robert G. Meyer. *"Analysis and Design of Analog Integrated Circuits"*. John Wiley & Sons, Inc., NY, second edition, 1995.

[25] E. Heaney, F. McGrath, P.O'Sullivan, and C Kermanrec. "Ultra Low Power Low Noise Amplifier for Wireless Communications". *GaAs IC Symposium*, pages 49–51, 1993.

[26] Qiuting Huang, Paolo Orsati, and Francesco Piazza. "Broadband, $0.25\mu m$ CMOS LNAs with Sub-2dB NF for GSM Application". *Proceedings of IEEE 1998 Custom Integrated Circuits Conference*, pages 67–70, 1998.

[27] Qiuting Huang and Robert Rogenmoser. "Speed Optimization of Edge-Triggered CMOS Circuits for Gigahertz Single-Phase Clocks". *IEEE Journal of Solid-State Circuits*, 31(3):456–465, 1996.

[28] Christopher Dennis Hull and Joo Leong Tham. "A Direct-Conversion Receiver for 900MHz (ISM Band) Spread-Spectrum Digital Cordless Telephone". *IEEE Journal of Solid-State Circuits*, 31(12):1955–1963, 1996.

[29] Paul J. Hurst, Roger A. Levinson, and David J. Block. "A Switched-Capacitor Delta-Sigma Modulator with Reduced Sensitivity to Op-Amp Gain". *IEEE Journal of Solid-State Circuits*, 28(6):691–696, 1993.

[30] Leon W. Couch II. *Digital and Analog Communication systems*. Prentice-Hall International, Inc., NY, fifth edition, 1987.

[31] Mohammed Ismail and Terri Fiez. *Analog VLSI Signal and Information Processing*. McGraw-Hill, Inc., New York, first edition, 1994.

[32] Arun Jayaraman and et al. "A Fully Integrated Broadband Direct-Conversion Receiver for DBS Applications". *2000 IEEE International Solid-State Circuits Conference Digest of Technical Papers*, pages 138–139, 2000.

[33] Hong jh Cong, John M. Andrews, and et al. "Multigigaherz CMOS Dual-Modulus Prescaler IC". *IEEE Journal of Solid-State Circuits*, 23(5):1189–1194, 1988.

[34] David A. Johns and Ken Martin. *Analog Integrated Circuit Design*. John Wiley & Sons, Inc., NY, first edition, 1997.

[35] Yuihi Kado, Masao Suzuki, and et al. "A 1GHz/0.9mW CMOS/SIMOX Divide-by-128/129 Dual-Modulus Prescaler Using a Divide-by-2/3 Synchronous Counter". *IEEE Journal of Solid-State Circuits*, 28(4):513–517, 1993.

[36] S. Kourtis. "W-CDMA: Aspects of Implementation". *IEEE Colloquium on UMTS - The R&D Challenges*, pages 10/1–10/8, 1998.

[37] S.-U Kwak, B-S. Song, and K. Bacrania. "A15-b 5-Msamples/s Low-Spurious CMOS ADC". *IEEE Journal of Solid-State Circuits*, 32(12):1866–1875, 1997.

[38] S.-H. Lee and B.-S. Song. "Digital-Domain Calibration of Multistep Analog-to-Digital Converters". *IEEE Journal of Solid-State Circuit*, 27(12):1679–1688, 1992.

[39] Xiaopeng Li and Mohammed Ismail. "Fast Simulation of Sigma-Delta Modulators using SPICE". *IEEE Circuits and Devices Magazine*, 15:7–9, January 1999.

[40] Xiaopeng Li and Mohammed Ismail. "A Single-chip CMOS Front-end Receiver Architecture for Multi-standard Wireless Applications". *Proceedings of 2001 IEEE International Symposium on Circuits and Systems*, 4:374–377, May 2001.

[41] Xiaopeng Li, H. Olsson, and Mohammed Ismail. "A High Speed Low Power Architecture for GHz CMOS Dual-Modulus Prescaler". *2000 IEEJ 4th Analog VLSI Workshop*, pages 6–9, 2000.

[42] Xiaopeng Li, Hong sun Kim, and Mohammed Ismail. "A Novel Design Approach for GHz CMOS Low Noise Amplifiers". *Proceedings of 1999 IEEE Radio and Wireless Conference*, pages 285–288, August 1999.

[43] Y.-M. Lin, B. Kim, and P. Gray. "A 13-b 2.5-MHz Self-Calibrated Pipelined A/D Converter in 3-mm CMOS ". *IEEE Journal of Solid-State Circuit*, 26(4):628–636, 1991.

[44] Bengt Lindoff. "Using a direct conversion receiver in edge terminals - a new DC offset compensation algorithm ". *Proceedings of the 11th IEEE International Symposium on Personal, Indoor and Mobile Radio Communications*, 2:956–963, 2000.

[45] Ting-Ping Liu. "5GHz CMOS Radio Transceiver Front-End Chipset". *2000 IEEE Internationa Solid-State Circuits conference Digest of Technical Papers*, pages 320–321, 2000.

[46] Mohammad Madihian, Tomislav Drenski, and et al. "A 5-GHz-Band Multifunctional BiCMOS Transceiver Chip for GMSK Modulation Wireless Systems". *IEEE Journal of Solid-State Circuits*, 34(1):25–32, January 1999.

[47] Tajinder Manku. "Microwave CMOS-devices and Circuits". *Proceedings of the IEEE 1998 Custom Integrated Circuits Conference*, pages 59–66, 1998.

[48] Augusto Manuel Marques and et al. "A 15-b Resolution 2-MHz Nyquist Rate Sigma-Delta ADC in a 1-um CMOS Technology". *IEEE Journal of Solid-State Circuits*, 33(7):1065–1075, July 1998.

[49] Ken martin and Adel S. Sedra. "Effects of the Op amp finite Gain and Bandwidth on the Performance of Switched-Capacitor Filters". *IEEE Transactions on Circuits and System*, CAS-28(8):1298–1308, August 1981.

[50] Sven Mattisson. "Architecture and Technology for Multi-standard Transceivers". *Proceedings of IEEE BCTM*, pages 82–85, 2001.

[51] F. Medeiro, B. Perez-Verdu, and et al. "Modeling OpAmp-Induced Harmonic Distortion for Switched-Capacitor $\Delta\Sigma$ Modulator Design". *Proceedings of the 1994 IEEE International Symposium on Circuits and Systems*, 5:445–448, 1994.

[52] Fernanado Medeiro and et al. "A 13bit, 2.2MS/s, 55mW Multibit Cascade Sigm-delta Modulator in CMOS 0.7um Single-Poly Technology". *IEEE Journal Solid-State Circuits*, 34(6):748–760, June 1999.

[53] Robert G. Meyer, William D. Mack, and Johannes J. EM. Hageraats. "A 2.5GHz BiCMOS transceiver for Wireless LAN's". *IEEE Journal of Solid-State Circuits*, 32(12):2097–2104, 1997.

[54] Murata. Filter and Switch Data Sheet.

[55] Paolo Orsatti, Francesco Piazza, and Qiuting Huang. "A 20mA-Receive, 55mA-Transmit, Single-Chip GSM Transceiver in 0.25um CMOS". *IEEE Journal of Solid-State Circuits*, 34(12):1869–1880, December 1999.

[56] Raj Pandya. *Mobile and Personal Communication Systems and Services*. IEEE Press, Inc., NJ, first edition, 2000.

[57] A. Parssinen, S. Lindfors, J. Ryynanen, S.I.Long, and K. Halonen. "1.8GH CMOS LNA with On-chip DC-coupling for a Subsampling Direct Conversion Front-end". *Proceedings of the 1998 IEEE International Symposium on Circuits and Systems*, 2:73–76, 1998.

[58] Aarmo Parssinen and et al. "A 2GHz Wide-Band Direct Conversion Receiver for WCDMA Applications". *IEEE Journal of Solid-State Circuits*, 34(12):1893–1903, 1999.

[59] Shahriar Rabii, Louis A. Williams III, Bernhard E. Boser, and Bruce A. Wooley. "MIDAS User Manual". *Version 3.1, Standford University*, 1997.

[60] Behzad Razavi. *Principles of Data Conversion System Design.* IEEE Press, NJ, first edition, 1995.

[61] Behzad Razavi. *RF Microelectronics*. Prentice Hall, NY, first edition, 1997.

[62] Arnold R.Feldman. "High-speed, Low-Power Sigma-Delta Modulators for RF Baseband Channel Applicaitons". *Ph.D theses, UC Berkeley*, 1998.

[63] J.C. Rudell, J.-J. Ou, T.B. Cho, G. Chien, F. Brianti, J.A. Weldon, and P.R. Gray. "A 1.9-GHz wide-band IF double conversion CMOS receiver for cordless telephone applications ". *IEEE Journal of Solid-State Circuits*, 32(12):2071–2088, December 1997.

[64] J. Ryynanen, K. Kivekas, J. Jussila, A. Parssinen, and K. Halonen. "A dual-band RF front-end for WCDMA and GSM applications". *IEEE Journal of Solid-State Circuits*, 36(8):1198–1204, 2001.

[65] S. Sampe and K. Feher. "Adaptive DC-offset compensation algorithm for burst mode operated direct conversion receivers ". *Proceedings of 42nd IEEE Vehicular Technology Conference*, 1:93–96, 1992.

[66] G. Schultes, P. Kreuzgruber, and A.L. Scholtz. "DECT Transceiver Architectures: Superheterodyne or direct conversion?". *Proceedings of 43st IEEE Vehicular Technology Conference*, pages 953–955, 1993.

[67] J. Sevenhans, A. Vanwelsenaers, and et al. "An integrated Si Bipolar RF Transceiver for a Zero IF 900 MHz GSM Digital Mobil Radio Front-End of a Hand Portable Phone". *Proceedings of the IEEE 1991 Custom Integrated Circuits Conference*, pages 7.7/1–7.7/4, 1991.

[68] Jan Sevenhans, Bart Verstraeten, and Silvio Taraborrelli. "Trends in Silicon Radio Large Scale Integration: Zero IF Receiver! Zero I&Q Transmitter! Zero Discrete Passives!". *IEEE Communicaitons Magazine*, pages 142–147, 2000.

[69] D. K. Shaeffer and T. H. Lee. "A 1.5V 1.5GHz CMOS Low Noise Amplifier". *IEEE J. Solid-State Circuits*, 32(5):745–759, May 1997.

[70] Arvin R. Shahani, Drek K. Shaeffer, and Thomas H. Lee. "A 12-mW Wide Dynamic Range CMOS Front-End for Portable GPS Receiver". *IEEE Journal of Solid-State Circuits*, 32(12):2061–2070, December 1997.

[71] Chunlei Shi, Xiaopeng Li, and Mohammed Ismail. "A Novel Nonlinear Settling Model of OTA for High-Resolution Switched-Capacitor Sigma-Delta Design". *Proceedings of 1999 International Conference on Microelectronics*, pages 22–24, 1999.

[72] Bluetooth SIG. *"Bluetooth Homepage"*. http://www.bluetooth.com, 2000.

[73] L. Singer, S. Ho, M. Timko, and D. Kelly. "A 12b 65Msample/s CMOS ADC with 82dB SFDR at 120MHz". *IEEE Solid-State Circuits Conference Digest of Technical Papers*, pages 38–39, February 2000.

[74] Richard Schreier Steven R. Norsworthy and Gabor C. Temes. *"Delta Sigma Data Converters Theory, Design and Simulation"*. IEEE Press, NJ, Second edition, 1997.

[75] John T. Stonick and Dongling Pan. "A New Robust Detector for Direct Conversion Receivers with DC Notching Filters". *Proceedings of the 11th IEEE International Symposium on Personal, Indoor and Mobile Radio Communications*, 2:956–963, 2000.

[76] John T. Stonick and Dongling Pan. "A New Robust Detector for Direct Conversion Receivers with DC Notching Filters". *Proceeding 2000 IEEE International Conference on Communications*, 1:387–391, 2000.

[77] Jon Strange and Simon Atkinson. "A Highly Integrated Radio Transceiver Chipset for DECT". *Proceedings of 1997 IEEE Radio Frequency Integrated Circuits Symposium*, pages 131–134, 1997.

[78] Hong sun Kim. "2.4GHz CMOS Receivers for Short-Range Wireless Applications". *Ph.D Dissertation, Ohio State University*, 2001.

[79] Hong sun Kim, Xiaopeng Li, , and M. Ismail. "A 2.4GHz CMOS Low Noise Amplifier using an Inter-stage Matching Inductor". *Proc. of 42nd Midwest Symposium on Circuits and Systems*, pages 1040–1043, August 1999.

[80] Chuc K. Tahanh, Stephen H. Lewis, and Paul J. Hurst. "A Second-Order Double-Sampled Delta-Sigma Modulator Using Individual-level Averaging". *IEEE Journal of Solid-State Circuits*, 32(8):1269–1273, 1997.

[81] Yiwu Tang, Adem Aktas, Mohammed Ismail, and Steven Bibyk. "A fully integrated dual-mode frequency synthesizer for GSM and wideband CDMA in 0.5um CMOS ". *Proceedings of IEEE Midwest Symposium on Circuits and Systems*, pages pp866–869, 2001.

[82] Gabor C. Temes. "Finite Amplifier Gain and Bandwidth Effects in Switched-Capacitor Filters". *IEEE Journal of Solid-State Circuits*, (3):358–361, 1980.

[83] R.J. Trihy and R.A. Rohrer. "A Switched Capacitor Circuit Simulator: AWEswit". *IEEE Journal of Solid-State Circuits*, 29(3):217–225, 1994.

[84] H. Tsurumi and T. Maeda. "Design study on a direct conversion receiver front-end for 280 MHz, 900 MHz, and 2.6 GHz band radio communication systems". *Proceedings of 41st IEEE Vehicular Technology Conference*, pages 457–462, 1991.

[85] W. Velt, J. Fenk, and et al. "A 2.7V 800MHz-2.1GHz Transceiver Chipset for Mobile Application in 25GHz f_t Si-Bipolar". *Proceedings of 1994 Bipolar/BiCMOS Circuits & Technology Meeting*, pages 175–178, 1994.

[86] Bing Wang, Hyuck M. kwon, and Jim Mitter. "Simple DC Removers for Digital FM Direct-conversion Receiver". *Proceedings of the 49th IEEE International Conference on Vehicular Technology*, 2:1222–1226, 2000.

[87] L. Williams and B. Wooley. "MIDAS-a functional simulator for mixed digital and analog sampled data systems". *Proc. IEEE 1992 Int. Symp. Circuits Syst.*, 5:2148–2151, May 1992.

[88] Ching-Yuan Yang, Cuan-Kaai Dehng, June-Ming Hsu, and Shen-Iuan Liu. "New Dynamic Flip-Flops for High-Speed Dual-Modulus Prescaler". *IEEE Journal of Solid-State Circuits*, 33(10):1568–1571, 1998.

[89] Jiren Yuan and Christer Svensson. "High-Speed CMOS Circuit Technique". *IEEE Journal of Solid-State Circuits*, 24(1):62–70, 1989.

Index

MIX
Papier aus verantwortungsvollen Quellen
Paper from responsible sources
FSC® C105338

If you have any concerns about our products,
you can contact us on
ProductSafety@springernature.com

In case Publisher is established outside the EU,
the EU authorized representative is:
Springer Nature Customer Service Center GmbH
Europaplatz 3, 69115 Heidelberg, Germany

Printed by Libri Plureos GmbH
in Hamburg, Germany